ADQI Consensus on AKI Biomarkers and Cardiorenal Syndromes

Contributions to Nephrology

Vol. 182

Series Editor

Claudio Ronco Vicenza

ADQI Consensus on AKI Biomarkers and Cardiorenal Syndromes

Volume Editors

Peter A. McCullough Novi, Mich.
John A. Kellum Pittsburgh, Pa.
Ravindra L. Mehta San Diego, Calif.
Patrick T. Murray Dublin
Claudio Ronco Vicenza

39 figures, 5 in color, and 21 tables, 2013

KARGER Basel · Freiburg · Paris · London · New York · New Delhi · Bangkok · Beijing · Tokyo · Kuala Lumpur · Singapore · Sydney

Contributions to Nephrology
(Founded 1975 by Geoffrey M. Berlyne)

Peter A. McCullough
Providence Park Heart Institute
Providence Park Hospital
47601 Grand River, B125
Novi, MI 48374 (USA)

John A. Kellum
Department of Critical Care Medicine
University of Pittsburgh
604, Scaife Hall
Pittsburgh, PA 15261 (USA)

Ravindra L. Mehta
Clinical Research
Department of Medicine
UCSD Medical Centre
200 W Arbor Drive 8342
San Diego, CA 92103 (USA)

Patrick T. Murray
UCD Catherine McAuley Centre
University College Dublin
Nelson Street
IRL-Dublin 7

Claudio Ronco
Department of Nephrology
Dialysis and Transplantation
International Renal Research Institute
San Bartolo Hospital
IT-36100 Vicenza

Library of Congress Cataloging-in-Publication Data

ADQI consensus on AKI biomarkers and cardiorenal syndromes / volume editors, Peter A. McCullough, John A. Kellum, Ravindra L. Mehta, Patrick T. Murray, Claudio Ronco.
p. ; cm. -- (Contributions to nephrology, ISSN 0302-5144 ; vol. 182)
Acute Dialysis Quality Initiative consensus on acute kidney injury biomarkers and cardiorenal syndromes
Includes bibliographical references and indexes.
ISBN 978-3-318-02406-7 (hard cover : alk. paper) -- ISBN 978-3-318-02407-4 (e-ISBN)
I. McCullough, Peter A., editor of compilation. II. Kellum, John A., editor of compilation. III. Mehta, Ravindra L., editor of compilation. IV. Murray, Patrick T., editor of compilation. V. Ronco, C. (Claudio), 1951- editor of compilation. VI. Acute Dialysis Quality Initiative. VII. Conference on Biomarkers in AKI (10th : 2011 : Dublin, Ireland) VIII. Conference on CRS (11th : 2012 : Venice, Italy) IX. Title: Acute Dialysis Quality Initiative consensus on acute kidney injury biomarkers and cardiorenal syndromes. X. Series: Contributions to nephrology ; v. 182. 0302-5144
[DNLM: 1. Acute Kidney Injury--Congresses. 2. Biological Markers--Congresses. 3. Cardio-Renal Syndrome--Congresses. W1 CO778UN v.182 2013 / WJ 342]
RD575
617.4'61059--dc23

2013010987

Bibliographic Indices. This publication is listed in bibliographic services, including Current Contents® and Index Medicus.

www.karger.com
Printed in Germany on acid-free and non-aging paper (ISO 9706) by Bosch Druck, Ergolding
ISSN 0302–5144
e-ISSN 1662–2782
ISBN 978–3–318–02406–7
e-ISBN 978–3–318–02407–4

Contents

Preface

There are several important topics in modern medicine but only few of them have realized significant advancement in recent years as acute kidney injury (AKI). This field is one of the most attractive and sparkling due to new information on pathophysiology and new interesting links to other medical disciplines such as intensive care, cardiac surgery and cardiology. In particular, heart-kidney interactions and their clinical consequences have been a subject of debate for many years but it is only recently that they have been subjected to a systematic appraisal and a true multidisciplinary approach. Linked to AKI and to cardiorenal syndromes (CRS), the fascinating world of new biomarkers represent a popular area of investigation and a new opportunity for advancement of therapy.

Biomarkers of AKI and the CRS represent a sort of moving target for research and investigators and require frequent reviews of the literature to make possible the transfer of all available information and solid evidence into clinical practice. This represents the main task of a consensus organization called ADQI (Acute Dialysis Quality Initiative). Born in 2000, ADQI has been the motor for development of consensus conferences, confrontation among experts and definition of current unanswered questions in the area of AKI, leading to new research pathways, new clinical trials and landmark studies.

Just to mention a few examples, we may recall the Best Kidney Study derived from ADQI 1 conference on CRRT, the ATN study based on the CRRT dose question posed by the ADQI group in the same conference, and further, the several RIFLE validation studies derived from the ADQI 2 conference dealing with definition of AKI and related research. After the above-mentioned conferences, many others took place over the years, always responding to the most urgent unanswered questions in the field of AKI.

This book contains the resolutions of the ADQI delegates obtained during the 10th conference on Biomarkers in AKI held in Dublin, September 2011, and the 11th conference on CRS held in Venice, December 2012.

The Dublin conference was organized to answer specific questions about new biomarkers and their use and utility in AKI. What are the most suitable candidate molecules and physiologic measures, how solid and evidence-based is the discovery phase? How can we incorporate the new biomarkers in the AKI conceptual model describing the evolution from susceptibility to insult, decreased GFR and organ death? Even if we have a positive biomarker pattern and can identify patients at risk or patients with early or even subclinical AKI, how is this information affecting our clinical behavior and practice? All these questions were posed to the panel of experts and four workgroups were created to establish current evidence and built statements and recommendations. The first part of the book is dedicated to the collection of workgroup products from the Dublin ADQI conference on biomarkers and therefore the reader should be able to receive from the relevant papers the most updated information.

The second part of the book is dedicated to the results of the Venice ADQI conference on CRS with a similar structure. A significant step forward had been made in 2008 when an ADQI conference was organized to discuss several aspects related to the CRS and a new definition/classification was established and vetted among nephrologists, cardiologists, and critical care specialists around the world. The new definition/classification has been the basis for new interactions among specialists and a new spirit of multidisciplinary collaboration. From that moment in fact, an important series of papers and a new body of research appeared in the literature, filling at least in part the gap of the past and spurring new interest in bridging the knowledge of cardiologists, nephrologists, and intensive care physicians. It was suggested that a modern approach to CRS should be multidisciplinary and patient-centered rather than specialty-oriented; the wide acceptance of the new definition classification represented the basis for further trials and cooperative research. The 2012 Venice ADQI conference has been dedicated to the appraisal of the current knowledge about the pathophysiological mechanisms involved in different forms of CRS. Thus, the second part of this book is designed to collect a series of contributions on the state-of-the-art knowledge and practice of CRS particularly focusing on the pathophysiology of the five subtypes. Acute and chronic mechanisms of damage are explored in depth with particular attention to the primacy of organ involvement and the subsequent pathways of organ crosstalk. After the executive summary of the conference, five contributions describe in detail each single subtype with its pathophysiological mechanisms, clinical features and possible therapeutic approaches. The possibility of an early recognition of the syndrome through novel diagnostic biomarkers represents and interesting integration and a link to the previous ADQI conference. CRS types 1 and 2 are typical conditions encountered in the field of cardiology and especially in the area of heart failure. Both acute heart failure leading to AKI and chronic heart failure leading to progressive renal insufficiency and chronic kidney disease represent conditions that may easily become interchangeable and sometimes it becomes very difficult to distinguish the two entities that often differ just for the time frame or the time window in the clinical history of the patient. Not only heart failure clinics are the stage for these disorders: acute heart failure may be induced by an acute coronary syndrome or by a sudden heart valve disease. Furthermore, CRS type 1 may be iatrogenically induced by an inappropriate use of diuretics and/or contrast media. So far, most heart failure trials excluded renal patients from enrollment and this has created a lack in knowledge for this category of patients. Therefore, CRS types 1 and 2 are frequently observed by cardiologists and a combined approach to these patients in conjunction with a nephrologist is strongly advised. CRS type 3 represents a totally different entity being a typical condition observed in critically ill patients admitted to the intensive care unit where AKI may induce a defective myocardial contractility and secondary heart failure. For this syndrome, the most appropriate case manager is probably the intensivist although a collegial evaluation of the patient and a multidisciplinary approach is recommended. CRS type 4 has been for many years defined as an increased cardiovascular risk in patients with chronic kidney disease. Today we know that even in the very early stages of chronic kidney disease, altered metabolism and gene expression may induce a complex derangement mediated by specific vitamin D receptors, oxidant stress, aldosterone receptors and so on. The fundamental role of the unfriendly milieu generated by uremia, hormonal dysregulation, and the combined effects of iron deficiency and anemia are explored in light of a significant combined damage. Further attention is placed in one contribution focusing on the delicate balance of organ crosstalk mediated by subclinical inflammation and a complicated network of mediators. The role of inflammation has been recently elucidated as an important mechanism for simultaneous heart and kidney damage and therefore sepsis has been taken as the prototype condition to study the pathophysiological mechanisms of CRS type 5.

Our intent as authors is to provide a compendium of the most recent research in the field of biomarkers, AKI and CRS. We hope the book will represent an important educational tool both for students and fellows, and for advanced investigators and clinical experts.

Peter A. McCullough, Novi, Mich.
John A. Kellum, Pittsburgh, Pa.
Ravindra L. Mehta, San Diego, Calif.
Patrick T. Murray, Dublin
Claudio Ronco, Vicenza

McCullough PA, Kellum JA, Mehta RL, Murray PT, Ronco C (eds): ADQI Consensus on AKI Biomarkers and Cardiorenal Syndromes. Contrib Nephrol. Basel, Karger, 2013, vol 182, pp 1–4
DOI: 10.1159/000349961

Acute Dialysis Quality Initiative (ADQI)

Claudio Ronco[a] · John A. Kellum[b] · Rinaldo Bellomo[c] · Ravindra L. Mehta[d]

[a]Department of Nephrology Dialysis and Transplantation & International Renal Research Institute (IRRIV), San Bortolo Hospital, Vicenza, Italy; [b]Clinical Research, Investigation, and Systems Modeling of Acute Illness Center, Department of Critical Care Medicine, University of Pittsburgh, Pittsburgh, Pa., USA; [c]Department of Intensive Care, Austin Hospital, Heidelberg, Vic., Australia; [d]Division of Nephrology and Hypertension, Department of Medicine, University of California San Diego, San Diego, Calif., USA

Abstract

ADQI – Acute Dialysis Quality Initiative – is a process that was created seeking consensus and evidence-based recommendations in the field of acute kidney injury. This organization gave birth to important classifications such as the RIFLE and the cardiorenal syndrome. Several conferences were organized in the last 10 years. Today, ADQI is a well-known initiative and it provides useful information to the physicians and nurses interested in acute kidney injury.

Acute kidney injury (AKI) is a common complication of critical illness and carries a high mortality. Indeed, in recent years it has become clear that AKI is an independent risk factor for death. Despite several attempts to improve treatment and to understand its pathogenesis, many aspects of this field remain the subject of controversy, confusion and lack of consensus. Until recently, there was no agreement even on the definition of AKI and there was no clear consensus as to how to prevent or manage it, including the choice, validity and relevance of animal models of AKI, the choice of appropriate physiological and clinical endpoints for trials of new treatments, the timing, intensity and modality of renal replacement therapy, the possible use of extracorporeal blood purification treatment for extrarenal diseases and the interaction between kidney dysfunction and abnormalities in other organs. Areas where consensus is lacking also include the principles that should guide fluid management in patients with AKI and the use of information technology to optimize all areas of patient care in this field.

Despite the fact that several guidelines and directions for appropriate medical management have been developed for end-stage renal disease, little had been done concerning AKI and its treatment. Only recently, the KDIGO guideline for AKI has been published but for many years the so-called evidence-based medicine (EBM) in the area of AKI was negligible. For these reasons, in the summer of 2000, a process was created seeking consensus and evidence-based recommendations in the field of AKI. The name of this process is ADQI – Acute Dialysis Quality Initiative.

The intent of ADQI is to provide an objective, dispassionate distillation of the literature and description of the current state of practice of dialysis and related therapies. The purpose is to develop a consensus of opinion, with evidence where possible, on best practice and to articulate a research agenda to focus on important unanswered questions. This approach is a blend of 'expert panel' and 'evidence appraisal' and was chosen in order to achieve the best of both methods.

The final objectives for ADQI are to: (i) standardize the process of diagnosis, prevention and treatment of AKI in the critically ill patient; (ii) develop consensus recommendations for best practice; (iii) facilitate the establishment of evidence-based guidelines for care where applicable, and (iv) identify questions for future research and consider study design options.

A standard methodology has been used for each of the eleven ADQI conferences to date [1]. Briefly, a group of experts comprising specialists of different disciplines but mostly nephrology and intensive care, gathered together for a two-day conference. Experts were assigned to small workgroups. Each workgroup was assigned a particular topic area. Topics were selected on the basis of the prevalence of the clinical problem, the variation in clinical practice, the potential influence on outcome, the potential for development of EBM guidelines and finally the availability of scientific evidence. Studies were identified via MEDLINE search, bibliographies of review articles and participants' files. Searches were limited to English language articles. However, articles written in other languages were used when identified and presented by members of the group. Evidence was classified according to levels per EBM methodology. Qualitative commentary was provided when deemed necessary by the group.

Each workgroup was composed of three to six members, one or two of whom served as the group facilitator. Summary statements were developed through a series of breakout sessions where individual workgroup members were required to identify key issues for which guidelines are needed and to classify the current state of consensus and identify supporting evidence for each issue. Workgroup members were then required to present their findings to the entire group, revising each statement as needed until a final version was agreed upon. The responsibilities for presenting the findings of the workgroup to the rest of the partici-

Table 1. ADQI conference topics

Topic	Location	Date
I. CRRT	New York, N.Y., USA	August 2000
II. ARF Research	Vicenza, Italy	May 2002
III. Non-Renal Blood Purification	Miami, Fla., USA	January 2003
IV. Prevention of ARF	Vicenza, Italy	May 2004
V. Fluid Management in AKI	Tambor, Costa Rica	January 2007
VI. AKI in Cardiac Surgery	Vicenza, Italy	May 2007
VII. Cardiorenal Syndrome	Venice, Italy	September 2008
VIII. Hepatorenal Syndrome	Kauai, Hawaii, USA	March 2010
IX. Toxicology	Denver, Colo., USA	June 2010
X. Biomarkers for AKI	Dublin, Ireland	September 2011
XI. Cardiorenal Syndrome 2	Venice, Italy	November 2012

pants were shared by each member on a rotating basis. Group facilitators revised workgroup findings as needed after each plenary session. Directives for future research were achieved by asking the participants to: (a) identify deficiencies in the literature, (b) determine if more evidence is necessary, and (c) if more evidence is necessary, articulate general research questions. When possible, pertinent study design issues were also considered.

Conference activities were divided into three steps: pre-conference, conference and post-conference. In the pre-conference step, workgroups were assembled and assigned to specific topics. Each group identified a list of key questions, conducted a systematic literature search and generated a bibliography of key studies. During the next step, the conference itself, the methodology was approved by the group and the conference was divided into breakout sessions where workgroups addressed the issues in their assigned topic area, and plenary sessions where their findings were presented, debated and refined. During the first plenary session, the key questions were discussed and debated. Revised versions of each question were then presented at the second plenary. At this point, evidence was assembled for each question and summary statements were drafted. These statements were further refined in subsequent plenary sessions until final versions were agreed upon. A writing committee assembled the individual reports from the workgroups. Each report was edited to conform to a uniform style and for length. The final reports were posted on the internet (www.ADQI.org) and mailed to each participant for comment and revision. Once final reports were completed, the writing committee summarized the individual reports into a final conference document to be published [2–9]. Table 1 lists the first eleven ADQI conferences and the topics covered.

The last two ADQI conferences were held in Dublin (Biomarkers of AKI) and Venice in 2012 (Cardiorenal Syndromes: Mechanisms and Pathophysiology).

The two recent topics overlap in many areas since the concept of acute organ damage and acute dysfunction can be applied to the kidney but also the heart and the lungs. A specific analysis of the pathophysiological mechanisms involved in the pathogenesis of an acute insult, progression from damage to dysfunction, maladaptive repair and progression of chronic organ disease has been undertaken with particular attention to the role of new biomarkers in the definition and classification of disease, its severity and staging.

It is our hope that the ADQI process will continue to blaze a trail in the wilderness and through the tireless efforts of its growing membership we will continue to bring about change for the better in the field of critical care nephrology. As our recent work has included the involvement of cardiologists, hepatologists and toxicologists, ADQI continues to be a multidisciplinary effort. From the proposed definitions of AKI (RIFLE), cardiorenal syndrome and hepatorenal disease, to clinical recommendations for use of CRRT and fluids that have served as the basis for clinical practice guidelines, ADQI has significantly impacted the way medicine is practiced around the world.

References

1 Kellum JA: The Acute Dialysis Quality Initiative: methodology. Adv Renal Replace Ther 2002;9:245–247.

2 Kellum JA, Mehta R, Ronco C: Acute Dialysis Quality Initiative (ADQI). Contrib Nephrol. Basel, Karger, 2001, vol 132, pp 258–265.

3 Ronco C, Kellum JA, Mehta R, Bellomo R, Palevsky PM: Introduction: Acute Dialysis Quality Initiative. Adv Renal Replace Ther 2002;9:227–228.

4 Ronco C, Kellum J, Mehta R: The Acute Dialysis Quality Initiative: The New York Conference. Adv Renal Replace Ther 2002;9:248–251.

5 Bellomo R, Kellum JA, Mehta R, Palevsky P, Ronco C: The Acute Dialysis Initiative II: The Vicenza Conference. Adv Renal Replace Ther 2002;9:290–293.

6 Clark WR, Paganini E, Weinstein D, Bartlett R, Sheinfeld G, Ronco C: Extracorporeal ultrafiltration for acute exacerbations of chronic heart failure: report from the Acute Dialysis Quality Initiative. Int J Artif Organs 2005;28:466–476.

7 Ronco C: Acute Dialysis Quality Initiative (ADQI): The PASSPORT Project. Int J Artif Organs 2005;28:438–440.

8 Ronco C, Bellomo R, Kellum JA: Acute Dialysis Quality Initiative: further steps toward improved practice in acute kidney injury. Int J Artif Organs 2008;31:89.

9 Ronco C, McCullough P, Anker SD, Anand I, Aspromonte N, Bagshaw SM, Bellomo R, Berl T, Bobek I, Cruz DN, Daliento L, Davenport A, Haapio M, Hillege H, House AA, Katz N, Maisel A, Mankad S, Zanco P, Mebazaa A, Palazzuoli A, Ronco F, Shaw A, Sheinfeld G, Soni S, Vescovo G, Zamperetti N, Ponikowski P, for the Acute Dialysis Quality Initiative (ADQI) Consensus Group: Cardiorenal syndromes: report from the consensus conference of the Acute Dialysis Quality Initiative. Eur Heart J 2010;31:703–711.

Claudio Ronco
Department of Nephrology, Dialysis and Transplantation and the International Renal Research Institute (IRRIV), San Bortolo Hospital
Viale Rodolfi 37, I-36100 Vicenza (Italy)
E-Mail cronco@goldnet.it

McCullough PA, Kellum JA, Mehta RL, Murray PT, Ronco C (eds): ADQI Consensus on AKI Biomarkers and Cardiorenal Syndromes. Contrib Nephrol. Basel, Karger, 2013, vol 182, pp 5–12
DOI: 10.1159/000349962

Implementation of Novel Biomarkers in the Diagnosis, Prognosis, and Management of Acute Kidney Injury: Executive Summary from the Tenth Consensus Conference of the Acute Dialysis Quality Initiative (ADQI)

Peter A. McCullough[1] · Josee Bouchard[2] · Sushrut S. Waikar[3] · Edward D. Siew[4] · Zoltan H. Endre[5] · Stuart L. Goldstein[6] · Jay L. Koyner[7] · Etienne Macedo[8] · Kent Doi[9] · Salvatore Di Somma[10] · Andrew Lewington[11] · Ravi Thadhani[12] · Raj Chakravarthi[13] · Can Ice[14] · Mark D. Okusa[15] · Jacques Duranteau[16] · Peter Doran[17] · Li Yang[18] · Bertrand L. Jaber[19] · Shane Meehan[20] · John A. Kellum[21] · Michael Haase[22] · Patrick T. Murray[23] · Dinna Cruz[24] · Alan Maisel[25] · Sean M. Bagshaw[26] · Lakhmir S. Chawla[27] · Ravindra L. Mehta[28] · Andrew D. Shaw[29] · Claudio Ronco[30] · for the Acute Dialysis Quality Initiative (ADQI) Consensus Group

[1]St. John Providence Health System, Warren, Mich., Providence Hospitals and Medical Centers, Southfield and Novi, Mich., St. John Macomb Oakland Center, Madison Heights, Mich., St. John Hospital and Medical Center, Detroit, Mich., USA; [2]Division of Nephrology and Hypertension, Division of Nephrology, Université de Montréal, Montréal, Que., Canada; [3]Renal Division, Department of Medicine, Brigham and Women's Hospital, Harvard Medical School, Boston, Mass., USA; [4]Division of Nephrology and Hypertension, Department of Medicine, Vanderbilt University School of Medicine, Nashville, Tenn., USA; [5]Christchurch Kidney Research Group, Department of Medicine, University of Otago, Christchurch, New Zealand; [6]Center for Acute Care Nephrology, Division of Nephrology and Hypertension, The Heart Institute, Cincinnati Children's Hospital Medical Center, Cincinnati, Ohio, USA; [7]Section of Nephrology, Department of Medicine, University of Chicago, Pritzker School of Medicine, Chicago, Ill., USA; [8]Division of Nephrology, University of São Paulo, São Paulo, Brazil; [9]Department of Nephrology and Endocrinology, The University of Tokyo, Tokyo, Japan; [10]Sant'Andrea Hospital, Second Faculty Medical School, La Sapienza University of Rome, Rome, Italy; [11]Leeds Teaching Hospitals, Leeds, UK; [12]Division of Nephrology, Massachusetts General Hospital, Boston, Mass., USA; [13]Division of Nephrology, Albany Medical College, Albany, N.Y., USA; [14]Department of Intensive Care Adults, Erasmus MC University Medical Centre Rotterdam,

Rotterdam, The Netherlands; [15]University of Virginia School of Medicine, Charlottesville, Va., USA; [16]Nephrology Unit, Le Kremlin Bicetre, Bicetre, France; [17]Medizinische Poliklinik, University of Munich, Munich, Germany; [18]Renal Division, Brigham and Women's Hospital, Boston, Mass., USA; [19]Division of Nephrology, Tufts Medical Center, Boston, Mass., USA; [20]Department of Pathology, Section of Nephrology, University of Chicago, Chicago, Ill., USA; [21]Clinical Research, Investigation, and Systems Modeling of Acute Illness Center, Department of Critical Care Medicine, University of Pittsburgh, Pittsburgh, Pa., USA; [22]Department of Nephrology, Hypertension, Diabetes and Endocrinology, Otto von Guericke University, Magdeburg, Germany; [23]School of Medicine and Medical Science, University College Dublin, Dublin, Ireland; [24]Department of Nephrology Dialysis and Transplantation, San Bortolo Hospital, Vicenza, Italy; [25]Department of Medicine and Cardiology, San Diego VA Medical Center and University of California, San Diego, Calif., USA; [26]Division of Critical Care Medicine, Faculty of Medicine and Dentistry, University of Alberta, Edmonton, Alta., Canada; [27]Department of Anesthesiology and Critical Care Medicine and Division of Renal Diseases and Hypertension, Department of Medicine, George Washington University Medical Center, Washington, D.C., USA; [28]Division of Nephrology and Hypertension, Department of Medicine, University of California San Diego, San Diego, Calif., USA; [29]Department of Anesthesiology, Duke University Medical Center, Durham, N.C., USA; [30]Department of Nephrology, International Renal Research Institute (IRRIV), San Bortolo Hospital, Vicenza, Italy

Abstract

Detection of acute kidney injury is undergoing a dynamic revolution of biomarker technology allowing greater, earlier, and more accurate determination of diagnosis, prognosis, and with powerful implication for management. Biomarkers can be broadly considered as any measurable biologic entity or process that allows differentiation between normal function and injury or disease. The ADQI (Acute Dialysis Quality Initiative) had its Ninth Consensus Conference dedicated to synthesis and formulation of the existing literature on biomarkers for the detection of acute kidney injury in a variety of settings. In the papers that accompany this summary, ADQI workgroups fully develop key concepts from a summary of the literature in the domains of early diagnosis, differential diagnosis, prognosis and management, and concurrent physiologic and imaging measures.

Acute kidney injury (AKI) is increasingly recognized as an in-hospital complication that is common and strongly related to increased hospital length of stay, use of intensive care unit services including hemodynamic monitoring, need for renal replacement therapy, rehospitalization, and death. When patients are followed after an episode of AKI, there are higher risks for the rapid progression of chronic kidney disease leading to end-stage renal disease requiring permanent dialysis therapy or renal transplantation. Hence, there is a great need to better understand AKI with the intent of developing prevention and treatment strategies. Prior to the past few years, the principle clinical tools used to detect AKI were serial measurement of serum creatinine (Cr), blood urea nitrogen, assessment of the urinalysis, and measurement of urine output. Agreement from multiple consensus groups (Acute Kidney Injury Network – AKIN, Risk, Injury, Failure, End-Stage Kidney Disease – RIFLE, and Kidney Disease Improving Global Outcomes –

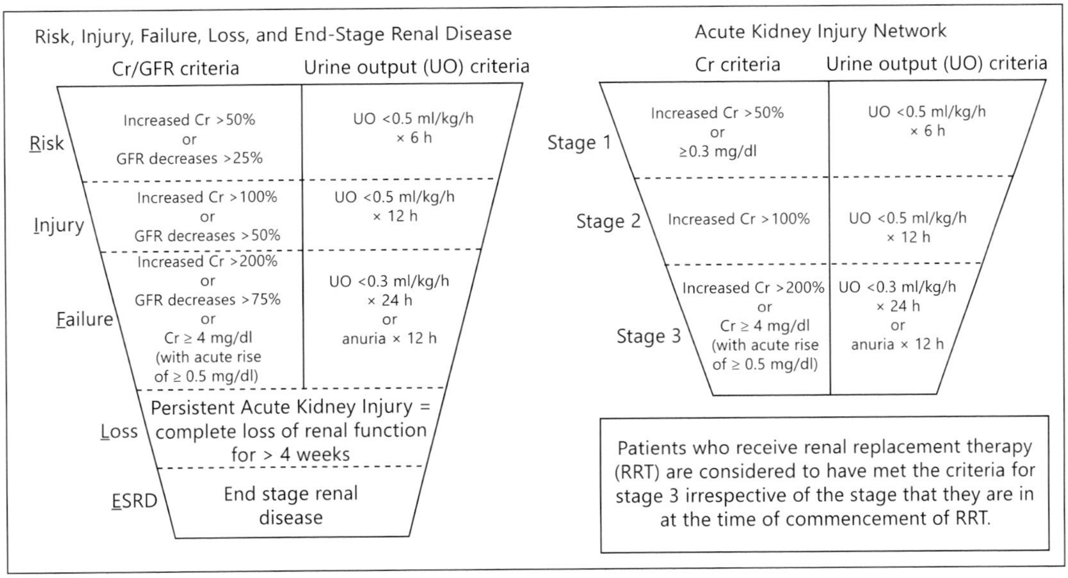

Fig. 1. The AKIN and RIFLE definitions and grading for AKI.

KDIGO) that AKI can be defined and graded has been a major advance in the phenotypic characterization as shown in figure 1 [1]. Thus the stage has been set for new technologies and measures to be assessed in the context of both the diagnosis and grading of AKI according to these standardized definitions [2].

Methods

The ADQI process was applied using previously described methodology taking advantage of key opinion leaders in the field and identifying appropriate topics for consensus [3]. The ADQI methodology comprises a systematic search for evidence with review and evaluation of relevant literature, establishment of clinical and physiologic outcomes for comparison of different treatments, description of current practice and analysis of areas in which evidence is lacking and a specific research agenda is required. ADQI activities were divided into a pre-conference, conference, and post-conference phase. Before the conference, topics were selected and workgroups assembled. Groups identified key questions and conducted a systematic literature search. During the conference, workgroups assembled in breakout sessions, as well as plenary sessions where their findings were presented, debated, and refined. Key questions were identified by the entire ADQI group, and the subgroups deliberated on these questions, bringing forth recommendations to the group as a whole. Deliberations followed three days of discussion amongst the attendees. Summary statements were then developed by the entire group and reported into the present document.

Results

Early Detection of Acute Kidney Injury

The first ADQI workgroup focused on the available literature on novel blood and urine biomarkers that have been tested in clinical studies for the early detection of AKI. Recognition that cells in the kidney produce and release measurable proteins in the setting of a variety of injuries has been an important step in the development of this field [4]. Injury appears to be as subtle as volume depletion which has been previously termed 'prerenal azotemia'. Data with novel markers suggest these episodes are indeed a form of AKI with some damage to cells and or complete nephron units [5]. More severe causes of renal damage include prolonged ischemia due to shock, sepsis and microcirculatory dysfunction, and direct toxicity due to injurious agents such as iodinated contrast, aminoglycosides, and vancomycin. Most of the markers mentioned in the accompanying article by McCullough and colleagues are either directly produced by renal tubular cells in response to upregulation of messenger RNA and protein synthesis, or are filtered proteins that have reduced reabsorption by the proximal tubular cells, and hence become measurable in greater quantities in the urine. Since the completion of the ADQI 10 meeting, and important discovery that tubular cell-cycle arrest markers (insulin-like growth factor-binding protein 7 (IGFBP7) and tissue inhibitor of metalloproteinases-2 (TIMP-2)) appear to be a brief (~36 h) harbinger to AKI that will develop in the next few days [6]. The concentrations of these proteins in the urine can be multiplied to create a value for use by clinicians to anticipate the immediate risk of AKI. The development of AKI from initiation of injury forward appears to be reflected by sharp elevation of neutrophil gelatinase-associated lipocalin (NGAL), kidney injury molecule-1 (KIM-1), L-type fatty acid-binding protein (L-FABP), and α-glutathione S-transferase (α-GST), π-GST, and interleukin-18 (IL-18). The preponderance of the data suggests these novel markers detect risk for AKI and probably actual AKI 24–48 h before the rise in serum Cr [7]. Importantly, there are cases where novel markers significantly elevate without a significant change in serum Cr or reduction in urine output, which is a late and ominous clinical development [8]. These cases of subclinical AKI appear to have important prognostic implications and probably should not be dismissed (fig. 2).

Differential Diagnosis

One of the most important uses of novel biomarkers is to help evaluate a patient who has a differential diagnosis in the setting of critical illness including pulmonary, heart, liver, and kidney failure. The novel markers in

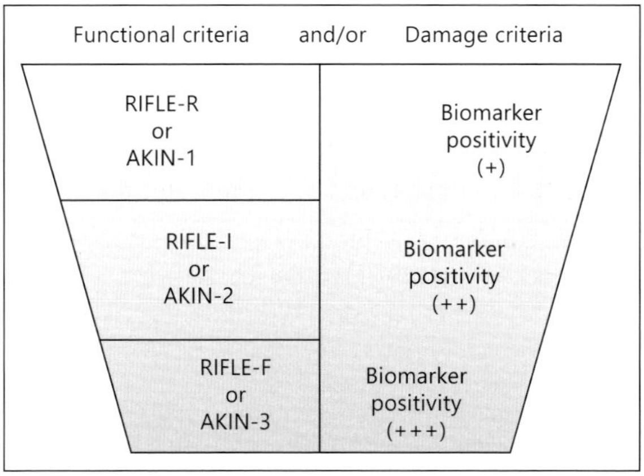

Fig. 2. Depiction of damage and functional markers in the detection of AKI. Reproduced with permission from ADQI [15].

aggregate appear to have moderate specificity for AKI. Some forms of injury, particularly those with abnormalities in cell signaling (cytokines, adhesion molecules) and microvascular dysfunction, appear to favor interleukins (IL-18), NGAL, and KIM-1 as relatively strong signals for AKI due to sepsis [9]. For most forms of injury, urinary levels of these markers are higher in concentration and have a greater rise than the same proteins measured in blood. When there are increases in urinary cystatin C and albumin from baseline, this suggests a failure in proximal renal tubular reabsorption [10]. This finding is complementary to information provided by the novel biomarkers above, particularly in establishing AKI in the differential diagnosis of the critically ill patient.

Prognosis and Management

Because AKI is strongly associated with early and late mortality, even when renal replacement therapy is offered, biomarkers indicating AKI are prognostic for mortality. In general, the greater the rise in a single marker, more consistent elevations of more than one marker, and when in conjunction with a reduction in glomerular filtration and urine output, there is a quantitative relationship between the severity of AKI and clinical outcomes. Thus, as shown in figure 3, with more prognostic information at an earlier stage of critical illness, there can be a series of clinical responses leading to reduced complications and possibly improvement in outcomes [1]. The underlying principles have now emerged: (1) novel markers allow early assessment of prognosis, (2) advanced warning can

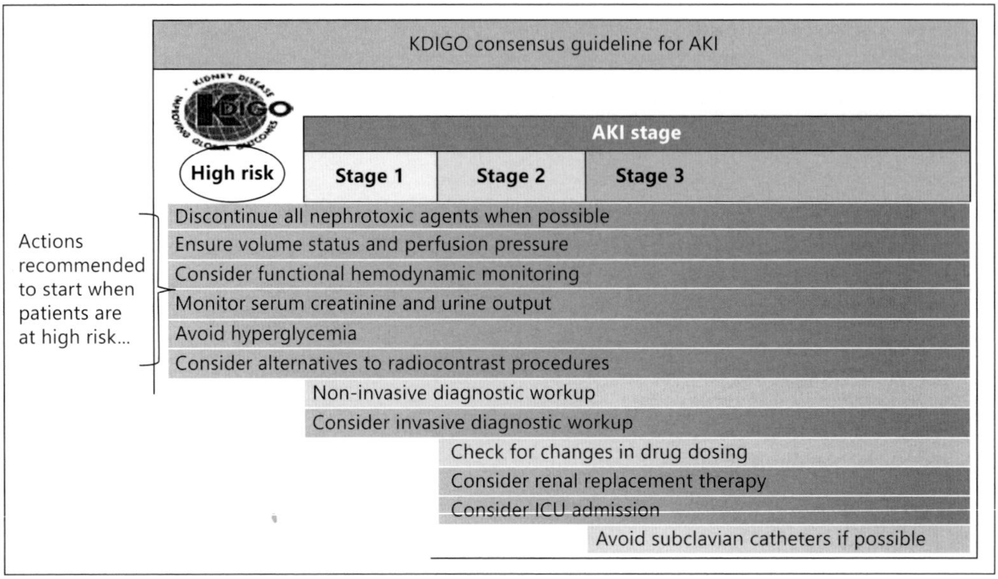

Fig. 3. Recognition of AKI and clinical responses. Adapted from KDIGO [1].

lead to changes in clinical processes, and (3) an alerted clinician and care team can more effectively care and communicate with patients and their families leading to improvements in delivery [11].

Physiologic and Imaging Measures

Cardiovascular medicine and oncology have taught us that the combination of laboratory measures and imaging/physiological assessments are powerful in terms of establishing internal validity, disease definitions, staging, and monitoring of progression of disease. Thus, in AKI, considerable work has been undertaken to understand core features of renal function including renal blood flow, regional renal perfusion, metabolism of glucose and other substrates, and the net control over renal filtration and urine output. A current standard of care in a case of de novo AKI includes renal ultrasonography which primarily establishes the presence of polycystic kidney disease, ureteral disease and obstruction, and helps somewhat to establish or support baseline chronic kidney disease with measurement of renal size. Ultrasound, however, is not helpful in diagnosing AKI in patient with normal size kidneys and no obstruction to urinary flow. The most promising novel approaches include renal time assessment of renal blood flow, oxygenation, and glomerular filtration [12, 13]. It is possible that discovery of a drop in renal filtration could occur many hours/days before serum Cr elevates, and thus give useful information to the clinician following the management principles discussed above [14].

Conclusions

AKI is a dynamic process that generates a multitude of biologic signals that when measured, become clinical biomarkers. Use of these markers appears to be valuable in the detection, differential diagnosis, prognosis, and management of patients in the setting of critical illness. Future research focused on the development and use of new diagnostic and therapeutic targets is critical to advance the field of critical care nephrology.

Acknowledgements

The ADQI 10 workgroup members are indebted to the independent monitoring of the ADQI process provided by the individuals listed below. These professionals provided content information and resources to the groups, but had no role in the writing or review of this or any of the ADQI 10 articles.

Funding Source: The 9th International ADQI Consensus Conference was funded in part by generous support in form of unrestricted educational grants from Alere Inc., and Astute Medical Inc. The sponsors had no input on the content of this paper.

Special Rotating Observers: Group A: Paul Clopton (USA), Joe Bonventre (USA); Group B: Paul Kimmel (USA), Bruce Molitoris (USA); Group C: Prasad Devarajan (USA), Maria Fitzgibbon (Ireland), and Group D: Eisei Noiri (Japan), Chirag Parikh (USA).

References

1 KDIGO AKI Guidelines: Summary of recommendation statements. Kidney Int Suppl 2012;2:8–12.

2 Uchino S, Bellomo R, Goldsmith D, Bates S, Ronco C: An assessment of the RIFLE criteria for acute renal failure in hospitalized patients. Crit Care Med 2006;34:1913–1917.

3 Kellum JA, Bellomo R, Ronco C: Acute Dialysis Quality Initiative (ADQI): methodology. Int J Artif Organs 2008;31:90–93.

4 Boudonck KJ, Rose DJ, Karoly ED, Lee DP, Lawton KA, Lapinskas PJ: Metabolomics for early detection of drug-induced kidney injury: review of the current status. Bioanalysis 2009;1:1645–1663.

5 McCullough PA, Chinnaiyan KM, Gallagher MJ, Colar JM, Geddes T, Gold JM, Trivax JE: Changes in renal markers and acute kidney injury after marathon running. Nephrology (Carlton) 2011;16:194–199.

6 Kashani K, Al-Khafaji A, Ardiles T, et al: Discovery and validation of cell cycle arrest biomarkers in human acute kidney injury. Critical Care 2013;17:R25.

7 Doi K, Negishi K, Ishizu T, Katagiri D, Fujita T, Matsubara T, et al: Evaluation of new acute kidney injury biomarkers in a mixed intensive care unit. Crit Care Med 2011;39:2464–2469.

8 Prowle JR, Liu YL, Licari E, Bagshaw SM, Egi M, Haase M, et al: Oliguria as predictive biomarker of acute kidney injury in critically ill patients. Crit Care 2011;15:R172.

9 Srisawat N, Kellum JA: Acute kidney injury: definition, epidemiology, and outcome. Curr Opin Crit Care 2011;17:548–555.

10 Urbschat A, Obermüller N, Haferkamp A: Biomarkers of kidney injury. Biomarkers 2011;16(suppl 1):S22–S30.

11 McCullough PA: Multimodality prevention of contrast-induced acute kidney injury. Am J Kidney Dis 2008;51:169–172.

12 Liu PS, Platt JF: CT angiography of the renal circulation. Radiol Clin North Am 2010;48: 347–365, viii–ix.

13 Inoue T, Matsuoka H, Higashi Y, Ueda S, Sata M, Shimada KE, Ishibashi Y, Node K, Vascular Failure Workshop Group: Flow-mediated vasodilation as a diagnostic modality for vascular failure. Hypertens Res 2008; 31:2105–2113.

14 Wang E, Meier DJ, Sandoval RM, Von Hendy-Willson VE, Pressler BM, Bunch RM, Alloosh M, Sturek MS, Schwartz GJ, Molitoris BA: A portable fiberoptic ratiometric fluorescence analyzer provides rapid point-of-care determination of glomerular filtration rate in large animals. Kidney Int 2012;81:112–117.

15 Acute Dialysis Quality Initiative. www.ADQI.org (accessed January 10, 2013).

Peter A. McCullough, MD, MPH
Providence Park Heart Institute, Providence Park Hospital
47601 Grand River, B125
Novi, MI 48374 (USA)
E-Mail peteramccullough@gmail.com

McCullough PA, Kellum JA, Mehta RL, Murray PT, Ronco C (eds): ADQI Consensus on AKI Biomarkers and Cardiorenal Syndromes. Contrib Nephrol. Basel, Karger, 2013, vol 182, pp 13–29
DOI: 10.1159/000349963

Diagnosis of Acute Kidney Injury Using Functional and Injury Biomarkers: Workgroup Statements from the Tenth Acute Dialysis Quality Initiative Consensus Conference

Peter A. McCullough[a] · Andrew D. Shaw[b] · Michael Haase[c] · Josee Bouchard[d] · Sushrut S. Waikar[e] · Edward D. Siew[f] · Patrick T. Murray[g] · Ravindra L. Mehta[h] · Claudio Ronco[i] · for the ADQI 10 Workgroup

[a]St. John Providence Health System, Detroit, Mich., USA; [b]Department of Anesthesiology, Duke University Medical Center, Durham, N.C., USA; [c]Department of Nephrology, Hypertension, Diabetology and Endocrinology, Otto von Guericke University, Magdeburg, Germany; [d]Division of Nephrology and Hypertension, Department of Medicine, University of Montreal, Montreal, Que., Canada; [e]Department of Medicine, Brigham and Women's Hospital, Harvard Medical School, Boston, Mass., USA; [f]Division of Nephrology and Hypertension, Department of Medicine, Vanderbilt University School of Medicine, Nashville, Tenn., USA; [g]Department of Medicine, University of Dublin, Dublin, Ireland; [h]Department of Medicine, University of California San Diego, San Diego, Calif., USA; [i]Department of Nephrology Dialysis and Transplantation & International Renal Research Institute, St. Bortolo Hospital, Vicenza, Italy

Abstract

Acute kidney injury (AKI) commonly occurs in hospitalized patients and is independently and strongly associates with morbidity and mortality. The clinical benefits of a timely and definitive diagnosis of AKI have not been fully realized due to limitations imposed by the use of serum creatinine and urine output to fulfill diagnostic criteria. These restrictions often lead to diagnostic delays, potential misclassification of actual injury status, and provide little information regarding underlying cause. Novel biomarkers of damage have shown ability to reflect ongoing kidney injury and help further refine existing Risk, Injury, Failure, Loss, End-stage kidney disease (RIFLE) and Acute Kidney Injury Network (AKIN) diagnostic criteria. A comprehensive review of the published literature to date was performed using previously published methodology of the Acute Dialysis Quality Initiative (ADQI) working group to es-

All authors contributed equally to this article.
ADQI 10 Workgroup members are listed in Appendix 1.

tablish consensus statements regarding (i) the overall implementation of injury biomarkers in the concept of AKI diagnosis, (ii) their clinical use, and (iii) future research. On the basis of published data on the ability of novel damage biomarkers to provide diagnostic and prognostic information on AKI, we recommend that novel damage biomarkers may, in the appropriate clinical setting and context (situation consistent with AKI), be used to diagnose AKI even in the absence of changes in serum creatinine or the presence of oliguria as described in the existing RIFLE/AKIN criteria for diagnosis of AKI. Adding injury biomarkers as a criterion for AKI will complement the ability of RIFLE/AKIN to define AKI. Promising diagnostic injury markers include neutrophil gelatinase-associated lipocalin (NGAL), kidney injury molecule 1 (KIM-1), interleukin 18 (IL-18) and liver-type fatty acid binding protein (L-FABP). However, there are currently insufficient data on damage biomarkers to support their use for AKI staging. Rigorous validation studies measuring the association between the novel damage biomarker(s) and clinically relevant outcomes are needed.

Acute kidney injury (AKI) is an increasingly common and potentially devastating complication in hospitalized individuals [1–3]. Because of the association between AKI and poor short- and long-term outcomes, new therapeutic and preventative strategies are desperately needed [4, 5]. The current diagnostic paradigm for AKI relies largely on biomarkers of glomerular filtration (serum creatinine (SCr), urine output (UO)) that have been in clinical use for over 50 years but are known to be insensitive and slow to change after kidney injury, often leading to a late and inaccurate diagnosis of AKI with resultant adverse outcomes [6–8].

Transcriptomic and proteomic techniques have identified several potential biomarkers of kidney injury including, but not limited to, neutrophil gelatinase-associated lipocalin (NGAL), kidney injury molecule 1 (KIM-1), interleukin 18 (IL-18) and liver-type fatty acid binding protein (L-FABP). This has led to the creation of a new field of research in AKI and potential application in clinical practice (fig. 1) [9]. The biology of these markers are varied and include functional (e.g. enzymatic), adaptive (e.g. inflammatory, siderophore), structural, or low molecular weight proteins normally filtered by the glomerulus and catabolized by healthy tubular epithelium such as cystatin C. In response to different types of injury, production of several of these markers is known to increase, and they may be released or actively secreted into the urine or circulation. Evidence to date suggests that several of these markers have diagnostic and prognostic value for patients at risk for or with AKI, and may help refine the current diagnostic approach to AKI [10].

Therefore, the current paradigm of AKI diagnosis needs to be reassessed in light of scientific advances in the understanding of the pathobiology of AKI facilitated in part by recent biomarker discovery programs. Considering the accumulating evidence for the utility of kidney damage biomarkers as early, sensitive, and prognostically useful diagnostic tests, members of the ADQI conference considered a reformulation of the diagnostic approach to AKI to include not only markers of function (e.g. changes in SCr and UO), but also markers of

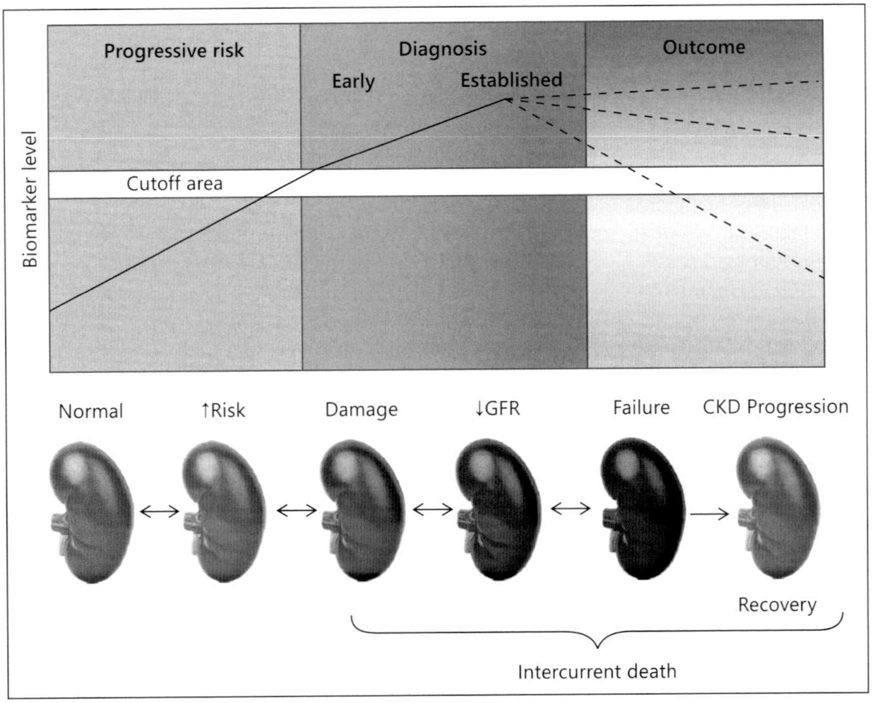

Fig. 1. AKI continuum and biomarker-based assessment. Reproduced with permission [50].

kidney damage without the need for changes in kidney function. The process included (i) the audit of overall implementation of kidney damage biomarkers in the diagnosis of AKI, (ii) their use in general versus selected patient cohorts and potential cut-off values for AKI diagnosis, and (iii) their ability for staging the clinical severity of AKI. The question of the need and opportunity for unifying functional and injury biomarkers for AKI diagnosis is discussed and recommendations are made on the basis of current knowledge.

Methods

We performed a systematic search and review of the available literature pre-conference, as described in detail elsewhere in this volume. Specifically we focused on the clinical value of renal damage biomarkers and their potential to be implemented in daily practice [11]. Studies were identified via Medline and Web of Science using the terms 'biomarker', and 'acute kidney injury (AKI)' combined with 'risk assessment', or '(early) diagnosis' or 'staging'. In view of the volume of the retrieved literature, only representative publications are cited in this review. Furthermore, we opted to limit this summary to clinical scenarios in which there was ample data. We did not include the specific clinical situations of renal or liver transplantation, or cirrhosis.

Table 1. Recognized risk factors and stressors for developing AKI adapted from Goldstein and Chawla [49]

Comorbidity	Kidney stressor
Age >65 years	Sepsis/shock/hypoxemia
Diabetes mellitus	Volume depletion
Cirrhosis/hepatic failure	Cardiopulmonary bypass
Heart failure	Exposure to nephrotoxins (e.g. vancomycin, gentamicin, NSAID, ACEI, ARB, cisplatin/carboplatin, methotrexate)
Chronic kidney disease	
eGFR <60 ml/min/1.73 m^2	
Albuminuria	Iodinated contrast
Glomerular diseases	Invasive mechanical ventilation
Renal transplant recipients	Multiorgan failure
Protein/calorie malnutrition	

eGFR = Estimated glomerular filtration rate; NSAID = nonsteroidal anti-inflammatory agent; ACEI = angiotensin converting enzyme inhibitor; ARB = angiotensin receptor blocker.

Results

Based on the literature identified prior to the conference, the following 7 key questions were considered: (1) Are conventional methods of risk assessment (prediction models, risk scores) adequate for calculating the risk for developing AKI? (2) Are current established diagnostic criteria for AKI adequate for a reliable and timely diagnosis of AKI? (3) Is there sufficient evidence for the inclusion of damage biomarkers as independent criteria for AKI diagnosis? (4) In which patients (at risk) should damage biomarkers be used for AKI diagnosis? (5) Which functional and damage biomarkers should be used for early AKI diagnosis? Is there sufficient evidence supporting cut-off values for damage biomarkers in AKI diagnosis? (6) Is there evidence of the optimal timing of a kidney damage biomarker measurement, and does it depend on the clinical scenario? (7) Is there sufficient evidence to permit a cost-benefit assessment for biomarker use, and what is the practical availability of damage biomarkers?

Risk Assessment

The risk for developing AKI is based on the cumulative effect of premorbid factors (e.g. comorbidities) that increase susceptibility to kidney damage and reduced function, and the type and severity of intercurrent processes (i.e. stressors) that impact the kidney (table 1). Several general and setting-specific prediction scores have been developed to identify which patients are at higher risk for

AKI following exposure to an acute renal stressor. The most commonly studied clinical settings are cardiac surgery (CABG), iodinated contrast and other nephrotoxin exposure, as well as in the critically ill. In most instances, the emerging biomarkers have not been assessed to predict susceptibility to AKI and are not incorporated in the scores. While risk assessment scores often perform well during development, few, if any, have demonstrated sufficient ability to be used clinically when generalized to external settings. Individualized risk prediction or early diagnosis is currently not possible from our review of the literature. Even when development-associated variables such as reduced kidney mass at birth and polymorphisms affecting the sympathetic and renin-angiotensin systems are included, accurate prediction is still not possible [12–16]. The most commonly studied clinical settings for AKI risk prediction are timed injury models (cardiac surgery, iodinated contrast exposure, other nephrotoxic agents) and untimed injury scenarios including critical illness with or without sepsis. These are briefly summarized below.

Cardiac Surgery

Three risk prediction scores have been recently reviewed and found to have an area under the receiver-operating characteristic curve (AROC) for renal replacement therapy (RRT) of roughly 0.79 [17]. These three scoring systems were not reproducible in other centers, were not suitable for estimation of the true incidence of postoperative need for RRT, were poorly calibrated and tended to underestimate the actual need for RRT [18]. These results are consistent with those from a prior study from the Mayo Clinic [19]. Many scores have not included timing of coronary angiography using iodinated contrast prior to CABG, which is an important determinant of AKI after the operation [20]. Other shortcomings included the use of intraoperative variables and variables that represent complications in the clinical course or are proxies for the baseline severity of illness.

Iodinated Contrast Exposure

The best developed prediction scores for contrast-induced AKI (CI-AKI) involve baseline renal function, contrast load, and measures of clinical course [21]. The most widely used and validated score was published by Mehran et al. [22] and utilizes baseline renal function (SCr or creatinine clearance), age, gender, hemoglobin, diabetes, history of heart failure, the development of shock requiring intra-aortic balloon counterpulsation, and the volume of contrast administered. This score can be calculated after a percutaneous coronary intervention and predicts the risk of CI-AKI and the need for RRT. In the validation subset of this study, there was only fair discrimination for those predicted to develop CI-AKI (C statistic = 0.67).

Nephrotoxin Exposure

A host of important therapeutic agents, such as aminoglycosides and cisplatin, can have direct nephrotoxicity [23, 24]. The ability to predict nephrotoxicity to stop a particular agent in order to reduce risk for AKI and improve patient safety is clinically relevant. There are known risk factors for aminoglycoside nephrotoxicity such as age, diabetes, baseline renal function, duration of therapy and frequency of dosing [25, 26]. However, to our knowledge, no clinical risk score has been developed in this setting.

Critical Illness

In critically ill patients admitted to intensive care units (ICU), Coritsidis et al. [27] have developed a clinical risk score including only serum albumin, presence or absence of sepsis and urine osmolality. This score outperformed the APACHE III score to predict AKI or mortality, but was not as predictive in critically ill patients in surgical ICUs [27]. Chawla et al. [28] also developed a clinical score to predict AKI in critically ill patients including many clinical variables. The only variables predictive of AKI were serum albumin concentration, increased A-a gradient, and presence of active cancer.

Sepsis

Hospitalized patients with multiple organ system dysfunction, infection, systemic inflammation, with or without obvious hemodynamic perturbations have the highest risk for AKI. Most attempts at risk prediction using clinical variables and or novel biomarkers have chosen in-hospital mortality as the outcome of interest and not AKI per se [29]. In addition, all risk prediction schemes use variables that in general are proxies for the severity of underlying illness and are not specific, independent predictors for AKI separate from mortality.

Diagnosis of AKI

The diagnosis of AKI using the AKIN (Acute Kidney Injury Network) or RIFLE (Risk, Injury, Failure, Loss, End-stage kidney disease) criteria is currently limited as SCr and UO can only indirectly reflect that kidney damage has occurred. These criteria, by definition, are retrospective in nature and require at least 6 h to several days, thus inherently delaying the clinical diagnosis of AKI [7, 30]. The UO criteria for AKI diagnosis (<0.5 ml/kg/h for >6 h) are hindered by limited sensitivity when diuretics are administered, reduced specificity in the presence of dehydration, and lack of practicality in measurement when an indwelling urinary catheter is not present [31]. These limitations have contributed to the inability to successfully characterize various aspects of the disease, consistently identify individuals at risk, and translate promising preclinical therapeutic strategies into successful treatments.

Over the last few years, several new markers have been developed to identify kidney injury, such as NGAL, KIM-1, IL-18 and L-FABP, among others [9]. Besides known filtration function markers such as SCr and cystatin C, there are filtered proteins that are increased due to glomerular injury (e.g. albumin), not reabsorbed at the tubular level due to tubular injury (albumin, cystatin C), markers that are released from damaged cells (α-glutathione S-transferase (GST), π-GST, collagen IV), or upregulated in response to cellular/tissue injury (NGAL, KIM-1, L-FABP, and IL-18). The availability of these new markers of 'kidney damage' afford a new opportunity to identify patients with AKI in addition to changes in SCr, glomerular filtration rate (GFR) or UO. As discussed in the introductory paper in this series and shown in figure 2, the combination of simultaneous assessment of a functional and damage markers can help stratify patients into 4 subgroups: no marker change, functional alone, damage alone, or change in both functional and damage markers. This categorization permits identification of a new category of patients who may have 'subclinical' AKI, i.e. an increase in damage markers without a simultaneous loss of kidney function (fig. 2, upper right quadrant). In these conditions, e.g. nephrotoxic drug exposure, loss of function may not develop at all or be seen at some time interval after detection of renal injury. Although direct evidence is sparse, it appears that these patients are at higher risk for adverse outcomes including need for RRT and mortality, than patients without an increase in damage biomarker levels [32].

Based on this new conceptual framework, we recommend a modification of the AKIN/RIFLE criteria to incorporate the use of kidney damage markers in addition to SCr, GFR and UO (fig. 3). Under this new diagnostic approach, in the appropriate clinical setting, AKI can be defined by abnormal levels of kidney injury biomarkers even in the absence of oliguria or elevated SCr. Our rationale for this proposal is based on an accumulating body of evidence validating various markers of kidney damage in many clinical conditions and populations where AKI constitutes a major problem. The bulk of evidence has focused on the following damage biomarkers including urinary/plasma NGAL, urinary KIM-1, urinary IL-18, and L-FABP. Several additional markers are in development and may also become commercially available over the short term. Most of these studies have focused on determining the time course of change in the biomarker in relation to a change in SCr. However, the timing of biomarker assessment has not been concurrent with SCr or UO assessment thereby introducing an inherent bias favoring the biomarkers. Nevertheless, the time course of the

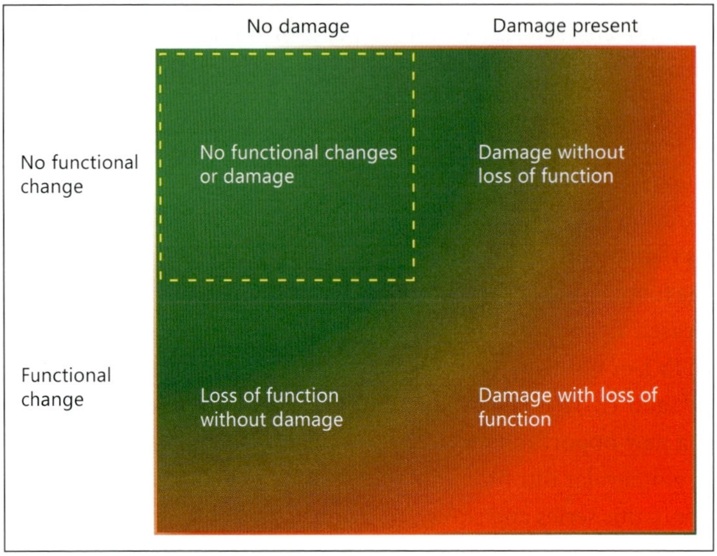

Fig. 2. New spectrum of AKI based on combination of functional and damage biomarkers. As illustrated, the combination of functional and damage biomarkers allows the clinician to differentiate a normal state of kidney function from abnormal to diagnose AKI. The current criteria for diagnosis include the lower two quadrants. This new spectrum enables the recognition of four subgroups of patients according to their AKI state. Patients negative for functional and kidney damage markers are considered to have no AKI (upper left quadrant). The ability to detect a state of damage alone (right upper quadrant) allows an expanded criteria for diagnosis of AKI. This may represent a 'subclinical' state in which loss of function might develop several days after detection of kidney damage or not at all, however, still associated with impaired outcomes. The bottom left quadrant indicates a dynamic change in renal filtration of SCr but without detectable kidney damage that may be physiologic such as seen in patients with dehydration. The right lower quadrant represents patients with functional and damage criteria of AKI associated with the worst prognosis. It is expected that the process is dynamic and patients will move from one phase to another during the course of their illness. Currently, there is limited information on what thresholds will be applicable for each of the damage biomarkers for the best diagnostic and staging criteria based on damage criteria alone. This will need to be defined in future studies. Reproduced with permission from ADQI [50].

biomarker change preceding any functional change suggests that it should be feasible to identify patients at a point in the course where they only have alteration in the injury markers. We have summarized the studies evaluating these markers for diagnosis of AKI below.

Cardiac Surgery

A consistent body of evidence spanning from ischemia/reperfusion-based animal experiments and numerous clinical studies linking damage biomarkers with

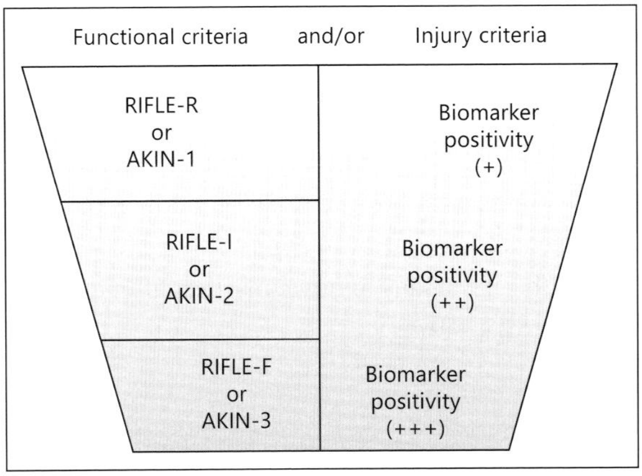

Fig. 3. New criteria for AKI diagnosis are displayed. In order to diagnose AKI selecting the worst criterion (function (RIFLE/AKIN) or damage) is recommended. In the appropriate clinical setting, this new damage biomarker criterion will enhance the ability of RIFLE/AKIN to define AKI. There are currently insufficient injury biomarker data to support staging of AKI, however AKI stages basing on renal function changes are suggested to remain. The semiquantitative trend for increasing biomarker severity associated with increasing kidney damage is suggested by the literature and is displayed by darkening background color as well as the symbols: +/++/+++. * Adapted from RIFLE/AKIN criteria. Reproduced with permission from ADQI [50].

AKI and adverse patient outcomes following cardiac surgery underlines their diagnostic value with the largest cumulative pediatric and adult patient cohort investigated using NGAL [9, 33].

Injury biomarkers for early diagnosis of AKI were investigated in more than 3,000 cardiac surgical patients with an AKI incidence of ~20% and have been shown to rise early, i.e. 2–12 h after surgery. The TRIBE (Translational Research Involving Biomarkers and Endpoints) study evaluated 1,219 adults undergoing cardiac surgery and reported that urine IL-18 and urine and plasma NGAL levels peaked within 6 h after surgery [9]. A clinical prediction model for AKI had an AROC of 0.69. Urine IL-18 and plasma NGAL significantly improved the AROC to 0.76 and 0.75, respectively. Urine IL-18 and plasma NGAL significantly improved risk prediction over the clinical models alone. The TRIBE investigators also evaluated 311 children undergoing surgery for congenital cardiac lesions using the same biomarkers. Similarly, they found that the novel damage markers rose within 6 h of surgery and that the AROC improved with urine IL-18 and urine NGAL, but not with plasma NGAL [34]. Thus, we believe novel markers in conjunction with the clinical evaluation and SCr can be used to diagnose AKI after cardiac surgery, in some cases before obvious renal failure becomes evident.

Iodinated Contrast Exposure

Kidney damage markers have been measured at baseline and serially after exposure (2, 4, 8, 24, and 48 h) to iodinated contrast and early injury could be demonstrated for both plasma and urinary NGAL and L-FABP prior to increases in SCr in several smaller studies in patients with normal renal function at baseline [35, 36]. Urinary IL-18 has also been shown to rise in concert with NGAL after contrast exposure [37]. In general, the elevations in novel markers after contrast exposure are less than those after cardiac surgery.

Nephrotoxin Exposure

Animal studies have suggested that urinary excretion of α-glutathione-S-transferase (α-GST), μ-GST, clusterin, and renal papillary antigen-1 could be used to monitor for cisplatin, gentamicin, or N-phenylanthranilic acid-associated nephrotoxicity [38]. In another animal study, KIM-1 appeared to detect gentamicin, mercury, and chromium associated-nephrotoxicity more readily than other markers including blood urea nitrogen (BUN), SCr, and NAG [39]. The US Food and Drug Administration and European Medicines Agency have qualified a panel of seven urinary kidney damage biomarkers that were shown in rodent studies to outperform SCr for the identification of nephrotoxicity as assessed by kidney histopathology [6, 40]. Thus, a negative or normal panel of kidney damage markers could be an advance in patient safety when performed on a regular basis with the use of common nephrotoxic agents (aminoglycosides, vancomycin, cisplatin, etc.). Clinical studies are clearly needed in this area and are being initiated in the USA through collaboration among several large pharmaceutical companies participating in the Predictive Safety Testing Consortium [40]. These studies will evaluate the performance of several injury biomarkers in patients exposed to cisplatin and aminoglycosides in two different populations.

Critical Illness

The published studies to date indicate that upon admission to an ICU or an emergency department, levels of multiple novel markers are predictive for the development of AKI, need for RRT, prolonged length of stay, and death [10, 41–44]. We believe that patients who are considered to be high risk for AKI with both risk factors and renal stressors on admission to the ICU (table 1), particularly those with sepsis, should be considered as candidates for testing with novel AKI biomarkers. In the absence of critical illness, the serial monitoring of novel markers for the early detection of AKI in sepsis is the current objective of several large, ongoing observational studies.

Normal Levels and Thresholds for Detection for Diagnosis of AKI

There may be biomarker changes within the 'normal range' that identify either increased risk or presence of early AKI (fig. 2), particularly in children. An NGAL cut-off value of <100 ng/ml (95–99% percentile), measured on standardized clinical laboratory platform seems to be useful for ruling out AKI in those children with normal baseline renal function [45]. At 2h after cardiac surgery, urine NGAL at the 100 ng/ml cutpoint had an AROC of 0.95, sensitivity of 82%, and specificity of 90% for detection of AKI later confirmed by a rise in SCr of ≥0.5 mg/dl. With NGAL (dynamic range 0–1,000 ng/ml), like many markers, the lower the value the greater the negative predictive value and conversely, the higher the value the greater the positive predictive value for the prediction of AKI later confirmed by SCr. Urine NGAL cut-offs may be somewhat lower compared to those for plasma NGAL, however, prospective studies measuring plasma and urine NGAL in the same patients at the same time are needed. The optimal cutpoint for NGAL in adult populations with baseline CKD risk factors or evident CKD is unknown. At the present time, in the absence of definitive data, it appears the optimal cutpoints for NGAL will be lowest in previously healthy pediatric populations undergoing cardiac surgery while in children with sepsis; cutpoints are likely to be higher given the untimed nature of AKI and the lack of baseline renal function information in many cases. In addition, for adults in the setting of cardiac surgery and CI-AKI, cutpoints will be higher than pediatric populations, but probably lower than those for adults in the setting of critical illness. Optimal cutpoints for the diagnosis of AKI need to be developed for L-FABP, KIM-1, and IL-18 and other emerging biomarkers measured in spot urine and serum. Progressively later generation assays may have different cut-off values and such values will always be manufacturer- and platform-dependent. The various assays and recommended cutpoints are presented elsewhere in this supplement.

Staging of AKI and the Role of Serial Measurement

Following diagnosis of AKI establishing the severity stage is essential to facilitate continued monitoring and prognostication assessment and monitoring and assess the prognosis. Currently, staging is based on the AKIN criteria utilizing changes in SCr and UO. As shown in figure 3, the proposed new criteria for damage would need to be refined further to determine which biomarkers and the thresholds that should be utilized for initial staging. Certainly this is an area for future research particularly to establish the thresholds for each biomarker and define the relationships of changes in different damage biomarkers to categorize severity. Additionally, a substantial gap of knowledge exists in the area determining the optimal times to measure biomarkers for staging AKI and to measure progression. Where the timing of the renal injurious event is known, the application of a single measurement of a damage biomarker or a panel has generally been performed several hours after the insult (cardiac surgery, contrast, nephrotoxins), but this has not been standardized. Since many of the novel markers are normally produced protein with constitutive and inducible rates of production, it is likely these markers continue to rise beyond the time windows measured in the studies to date.

Based on our current state of knowledge, the staging of AKI by kidney damage biomarkers alone cannot be justified, at present, since no discriminatory cut-off values to separate mild from moderate or severe AKI can be extracted from the literature. In addition, AKI is a clinical diagnosis which cannot be staged by biomarkers until we understand the time course and clinical outcomes associated with biomarker excursions. We understand these ADQI Consensus Conference recommendations/suggestions to be in line with the recently published recommendations of Kidney Disease: Improving Global Outcomes (KDIGO) where SCr and UO were recommended 'to remain the best functional biomarkers for AKI' at that time [46].

Serial measurements may allow observation of a progression of AKI (fig. 1). We were unable to evaluate or anticipate the cost-to-benefit ratio of serial measurements of novel damage biomarkers in the early diagnosis of AKI [47]. Thus, while we recommend the use of novel markers for AKI, we cannot recommend a serial testing schedule at this time.

Opportunities for Future Research

Further research will be needed to validate this updated classification. We expect that ongoing and future studies will introduce further improved biomarkers and refined cut-off values for staging using those investigated to date. Future clinical studies should also aim to test whether earlier diagnosis of AKI guided by injury biomarkers may translate into effective treatment and clinical benefits. The use and interpretation of markers outside of these clinical settings cannot be extrapolated with a high degree of confidence. Available clinical laboratory platforms for biomarker measurement will contribute to early analytical harmonization. Development of commercially available assays for other biomarkers will be necessary to optimize the benefits of research and clinical implementation in the field. Refinement of optimal cutpoints or magnitude of change will be key variables to ascertain and validate in marker research. Measuring SCr in the same sample as the novel marker and quantifying UO whenever feasible will yield to more consistent comparative analyses from multiple studies evaluating different markers. At some point in time, consideration may be given to one or more novel markers of AKI to become the new reference standard for AKI. In order for this to occur, it has been proposed that three principles be considered: (1) the consequences of the new tests should be understood through disagreements between the old and new tests; (2) resolving disagreements requires a fair umpire test, and (3) such an umpire test can be supportive clinical information (causal exposure, other testing, response to treatment) [48]. Considerable opportunities exist to address these principles in research of risk assessment and early detection of AKI in the appropriate 'at-risk' populations answering the following questions: (1) What are the consequences in patients in whom the novel markers of AKI disagree with the conventional AKIN or RIFLE classification schemes? (2) What could a 'fair umpire' test look like for AKI taking into consideration the clinical setting, causal exposure (cardiac surgery, contrast, infection, etc.), other tests, and response to treatment? (3) Are there genetic and early development factors that put an individual at increased risk for AKI after controlling for other known risk factors and stressors? (4) Are clinical decisions based on the use of novel biomarkers able to achieve improved clinical outcomes when compared with current methods of risk assessment and early detection of AKI? (5) Will a single or multiple marker approach be optimal in such an evaluation, and at one time point only or with serial measurements? (6) What are the thresholds for each novel marker in the setting of normal, abnormal, and unknown renal function at baseline? (7) What is the concordance or internal consistency of multiple markers in the setting of proven AKI? (8) Can injury markers alone identify 'preclinical AKI' before any conventional signs of AKI develop? (9) What are the costs and benefits associated with clinical strategies that utilize novel markers of AKI?

Conclusion

Understanding of baseline risk factors, renal stressors, and incorporation of the results of novel kidney damage markers (either assessed in blood or urine, or via other innovative physiological or imaging measures) may provide a step forward in avoiding harm to the kidney and improving patient safety and devise early preventive strategies. Several novel kidney damage biomarkers are now commercially available for AKI diagnosis, including blood and urine NGAL and L-FABP. We recognize that other markers are in late-stage clinical development and may in fact be superior to the ones currently available. We recommend that novel kidney damage biomarkers be integrated with the RIFLE/AKIN criteria for diagnosis of AKI. In the appropriate clinical setting, AKI may be diagnosed by either an abnormal damage biomarker OR changes in SCr or UO. This new damage criterion will complement the ability of RIFLE/AKIN to define AKI. There are insufficient data on damage biomarkers to support their use for staging of AKI. Carefully constructed clinical research to guide this field will be critical to determine whether the additional cost is worthwhile and to leverage outcomes for benefit in this population.

References

1 Uchino S, Bellomo R, Goldsmith D, Bates S, Ronco C: An assessment of the RIFLE criteria for acute renal failure in hospitalized patients. Crit Care Med 2006;34:1913–1917.

2 Xue JL, Daniels F, Star RA, Kimmel PL, Eggers PW, Molitoris BA, Himmelfarb J, Collins AJ: Incidence and mortality of acute renal failure in Medicare beneficiaries, 1992 to 2001. J Am Soc Nephrol 2006;17:1135–1142.

3 Waikar SS, Wald R, Chertow GM, Curhan GC, Winkelmayer WC, Liangos O, Sosa MA, Jaber BL: Validity of International Classification of Diseases, Ninth Revision, Clinical Modification Codes for Acute Renal Failure. J Am Soc Nephrol 2006;17:1688–1694.

4 Chertow GM, Burdick E, Honour M, Bonventre JV, Bates DW: Acute kidney injury, mortality, length of stay, and costs in hospitalized patients. J Am Soc Nephrol 2005;16: 3365–3370.

5 Wald R, Quinn RR, Luo J, Li P, Scales DC, Mamdani MM, Ray JG, University of Toronto Acute Kidney Injury Research Group: Chronic dialysis and death among survivors of acute kidney injury requiring dialysis. JAMA 2009;302:1179–1185.

6 Vaidya VS, Ozer JS, Dieterle F, Collings FB, Ramirez V, Troth S, Muniappa N, Thudium D, Gerhold D, Holder DJ, Bobadilla NA, Marrer E, Perentes E, Cordier A, Vonderscher J, Maurer G, Goering PL, Sistare FD, Bonventre JV: Kidney injury molecule-1 outperforms traditional biomarkers of kidney injury in preclinical biomarker qualification studies. Nat Biotechnol 2010;28:478–485.

7 Waikar SS, Bonventre JV: Creatinine kinetics and the definition of acute kidney injury. J Am Soc Nephrol 2009;20:672–679.

8 Shiao CC, Wu VC, Li WY, Lin YF, Hu FC, Young GH, Kuo CC, Kao TW, Huang DM, Chen YM, Tsai PR, Lin SL, Chou NK, Lin TH, Yeh YC, Wang CH, Chou A, Ko WJ, Wu KD, National Taiwan University Surgical Intensive Care Unit-Associated Renal Failure Study Group: Late initiation of renal replacement therapy is associated with worse outcomes in acute kidney injury after major abdominal surgery. Crit Care 2009;13:R171.

9 Parikh CR, Coca SG, Thiessen-Philbrook H, Shlipak MG, Koyner JL, Wang Z, Edelstein CL, Devarajan P, Patel UD, Zappitelli M, Krawczeski CD, Passik CS, Swaminathan M, Garg AX, for the TRIBE-AKI Consortium: Postoperative biomarkers predict acute kidney injury and poor outcomes after adult cardiac surgery. J Am Soc Nephrol 2011;22: 1748–1757.

10 Haase M, Bellomo R, Devarajan P, Schlattmann P, Haase-Fielitz A, NGAL Meta-Analysis Investigator Group: Accuracy of neutrophil gelatinase-associated lipocalin (NGAL) in diagnosis and prognosis in acute kidney injury: a systematic review and meta-analysis. Am J Kidney Dis 2009;54:1012–1024.

11 Kellum JA, Bellomo R, Ronco C: Acute Dialysis Quality Initiative (ADQI): methodology. Int J Artif Organs 2008;31:90–93.

12 Stafford-Smith M, Podgoreanu M, Swaminathan M, Phillips-Bute B, Mathew JP, Hauser EH, Winn MP, Milano C, Nielsen DM, Smith M, Morris R, Newman MF, Schwinn DA, Perioperative Genetics and Safety Outcomes Study (PEGASUS) Investigative Team: Association of genetic polymorphisms with risk of renal injury after coronary bypass graft surgery. Am J Kidney Dis 2005;45:519–530.

13 Du Cheyron D, Fradin S, Ramakers M, Terzi N, Guillotin D, Bouchet B, Daubin C, Charbonneau P: Angiotensin-converting enzyme insertion/deletion genetic polymorphism: its impact on renal function in critically ill patients. Crit Care Med 2008;36:3178–3183.
14 Haase-Fielitz A, Haase M, Bellomo R, Lambert G, Matalanis G, Story D, Doolan L, Buxton B, Gutteridge G, Luft FC, Schunck WH, Dragun D: Decreased catecholamine degradation associates with shock and kidney injury after cardiac surgery. J Am Soc Nephrol 2009;20:1393–1403.
15 Vehaskari VM: Prenatal programming of kidney disease. Curr Opin Pediatr 2010;22: 176–182.
16 McCullough PA, Lepor NE: The deadly triangle of anemia, renal insufficiency, and cardiovascular disease: implications for prognosis and treatment. Rev Cardiovasc Med 2005; 6:1–10.
17 Vives M, Monedero P, Perez-Valdivieso JR, Garcia-Fernandez N, Lavilla J, Herreros J, Bes-Rastrollo M, Spanish Renal Dysfunction in Cardiac Surgery Group: External validation and comparison of three scores to predict renal replacement therapy after cardiac surgery: a multicenter cohort. Int J Artif Organs 2011;34:329–338.
18 Heise D, Sundermann D, Braeuer A, Quintel M: Validation of a clinical score to determine the risk of acute renal failure after cardiac surgery. Eur J Cardiothorac Surg 2010;37: 710–716.
19 Englberger L, Suri RM, Li Z, Dearani JA, Park SJ, Sundt TM 3rd, Schaff HV: Validation of clinical scores predicting severe acute kidney injury after cardiac surgery. Am J Kidney Dis 2010;56:623–631.
20 Mehta RH, Honeycutt E, Patel UD, Lopes RD, Williams JB, Shaw LK, O'Brien SM, Califf RM, Hughes GC, Sketch MH Jr: Relationship of the time interval between cardiac catheterization and elective coronary artery bypass surgery with postprocedural acute kidney injury. Circulation 2011; 124(suppl):S149–S155.
21 McCullough PA, Adam A, Becker CR, Davidson C, Lameire N, Stacul F, Tumlin J, CIN Consensus Working Panel: Risk prediction of contrast-induced nephropathy. Am J Cardiol 2006;98:27K–36K.
22 Mehran R, Aymong ED, Nikolsky E, Lasic Z, Iakovou I, Fahy M, Mintz GS, Lansky AJ, Moses JW, Stone GW, Leon MB, Dangas G: A simple risk score for prediction of contrast-induced nephropathy after percutaneous coronary intervention: development and initial validation. J Am Coll Cardiol 2004;44:1393–1399.
23 Boudonck KJ, Rose DJ, Karoly ED, Lee DP, Lawton KA, Lapinskas PJ: Metabolomics for early detection of drug-induced kidney injury: review of the current status. Bioanalysis 2009;1:1645–1663.
24 Perazella MA, Moeckel GW: Nephrotoxicity from chemotherapeutic agents: clinical manifestations, pathobiology, and prevention/therapy. Semin Nephrol 2010;30:570–581.
25 Bertino JS Jr, Booker LA, Franck PA, Jenkins PL, Franck KR, Nafziger AN: Incidence of and significant risk factors for aminoglycoside-associated nephrotoxicity in patients dosed by using individualized pharmacokinetic monitoring. J Infect Dis 1993;167:173–179.
26 Munckhof WJ, Grayson ML, Turnidge JD: A meta-analysis of studies on the safety and efficacy of aminoglycosides given either once daily or as divided doses. J Antimicrob Chemother 1996;37:645–663.
27 Coritsidis GN, Guru K, Ward L, Bashir R, Feinfeld DA, Carvounis CP: Prediction of acute renal failure by 'bedside formula' in medical and surgical intensive care patients. Ren Fail 2000;22:235–244.
28 Chawla LS, Abell L, Mazhari R, Egan M, Kadambi N, Burke HB, Junker C, Seneff MG, Kimmel PL: Identifying critically ill patients at high risk for developing acute renal failure: a pilot study. Kidney Int 2005;68:2274–2280.
29 Doi K, Negishi K, Ishizu T, Katagiri D, Fujita T, Matsubara T, Yahagi N, Sugaya T, Noiri E: Evaluation of new acute kidney injury biomarkers in a mixed intensive care unit. Crit Care Med 2011.
30 Himmelfarb J, Ikizler TA: Acute kidney injury: changing lexicography, definitions, and epidemiology. Kidney Int 2007;71:971–976.
31 Prowle JR, Liu YL, Licari E, Bagshaw SM, Egi M, Haase M, Haase-Fielitz A, Kellum JA, Cruz D, Ronco C, Tsutsui K, Uchino S, Bellomo R: Oliguria as predictive biomarker of acute kidney injury in critically ill patients. Crit Care 2011;15:R172.

32 Haase M, Devarajan P, Haase-Fielitz A, Bellomo R, Cruz DN, Wagener G, Krawczeski CD, Koyner JL, Murray P, Zappitelli M, Goldstein SL, Makris K, Ronco C, Martensson J, Martling CR, Venge P, Siew E, Ware LB, Ikizler TA, Mertens PR: The outcome of neutrophil gelatinase-associated lipocalin-positive subclinical acute kidney injury: a multicenter pooled analysis of prospective studies. J Am Coll Cardiol 2011;57:1752–1761.
33 Krawczeski CD, Goldstein SL, Woo JG, Wang Y, Piyaphanee N, Ma Q, Bennett M, Devarajan P: Temporal relationship and predictive value of urinary acute kidney injury biomarkers after pediatric cardiopulmonary bypass. J Am Coll Cardiol 2011;58:2301–2309.
34 Parikh CR, Devarajan P, Zappitelli M, Sint K, Thiessen-Philbrook H, Li S, Kim RW, Koyner JL, Coca SG, Edelstein CL, Shlipak MG, Garg AX, Krawczeski CD, TRIBE-AKI Consortium: Postoperative biomarkers predict acute kidney injury and poor outcomes after pediatric cardiac surgery. J Am Soc Nephrol 2011;22:1737–1747.
35 Bachorzewska-Gajewska H, Malyszko J, Sitniewska E, Malyszko JS, Dobrzycki S: Neutrophil-gelatinase-associated lipocalin and renal function after percutaneous coronary interventions. Am J Nephrol 2006;26:287–292.
36 Bachorzewska-Gajewska H, Malyszko J, Sitniewska E, Malyszko JS, Poniatowski B, Pawlak K, Dobrzycki S: NGAL (neutrophil gelatinase-associated lipocalin) and cystatin C: are they good predictors of contrast nephropathy after percutaneous coronary interventions in patients with stable angina and normal serum creatinine? Int J Cardiol 2008;127:290–291.
37 Ling W, Zhaohui N, Ben H, Leyi G, Jianping L, Huili D, Jiaqi Q: Urinary IL-18 and NGAL as early predictive biomarkers in contrast-induced nephropathy after coronary angiography. Nephron Clin Pract 2008;108:c176–c181.
38 Harpur E, Ennulat D, Hoffman D, Betton G, Gautier JC, Riefke B, Bounous D, Schuster K, Beushausen S, Guffroy M, Shaw M, Lock E, Pettit S, HESI Committee on Biomarkers of Nephrotoxicity: Biological qualification of biomarkers of chemical-induced renal toxicity in two strains of male rat. Toxicol Sci 2011;122:235–252.
39 Zhou Y, Vaidya VS, Brown RP, Zhang J, Rosenzweig BA, Thompson KL, Miller TJ, Bonventre JV, Goering PL: Comparison of kidney injury molecule-1 and other nephrotoxicity biomarkers in urine and kidney following acute exposure to gentamicin, mercury, and chromium. Toxicol Sci 2008;101:159–170.
40 Dieterle F, Sistare F, Goodsaid F, Papaluca M, Ozer JS, Webb CP, Baer W, Senagore A, Schipper MJ, Vonderscher J, Sultana S, Gerhold DL, Phillips JA, Maurer G, Carl K, Laurie D, Harpur E, Sonee M, Ennulat D, Holder D, Andrews-Cleavenger D, Gu YZ, Thompson KL, Goering PL, Vidal JM, Abadie E, Maciulaitis R, Jacobson-Kram D, Defelice AF, Hausner EA, Blank M, Thompson A, Harlow P, Throckmorton D, Xiao S, Xu N, Taylor W, Vamvakas S, Flamion B, Lima BS, Kasper P, Pasanen M, Prasad K, Troth S, Bounous D, Robinson-Gravatt D, Betton G, Davis MA, Akunda J, McDuffie JE, Suter L, Obert L, Guffroy M, Pinches M, Jayadev S, Blomme EA, Beushausen SA, Barlow VG, Collins N, Waring J, Honor D, Snook S, Lee J, Rossi P, Walker E, Mattes W: Renal biomarker qualification submission: a dialog between the FDA-EMEA and Predictive Safety Testing Consortium. Nat Biotechnol 2010;28:455–462.
41 De Geus HR, Bakker J, Lesaffre EM, le Noble JL: Neutrophil gelatinase-associated lipocalin at ICU admission predicts for acute kidney injury in adult patients. Am J Respir Crit Care Med 2011;183:907–914.
42 Shapiro NI, Trzeciak S, Hollander JE, Birkhahn R, Otero R, Osborn TM, Moretti E, Nguyen HB, Gunnerson K, Milzman D, Gaieski DF, Goyal M, Cairns CB, Kupfer K, Lee SW, Rivers EP: The diagnostic accuracy of plasma neutrophil gelatinase-associated lipocalin in the prediction of acute kidney injury in emergency department patients with suspected sepsis. Ann Emerg Med 2010;56:52–59.e1.
43 Siew ED, Ware LB, Gebretsadik T, Shintani A, Moons KG, Wickersham N, Bossert F, Ikizler TA: Urine neutrophil gelatinase-associated lipocalin moderately predicts acute kidney injury in critically ill adults. J Am Soc Nephrol 2009;20:1823–1832.

44 Siew ED, Ikizler TA, Gebretsadik T, Shintani A, Wickersham N, Bossert F, Peterson JF, Parikh CR, May AK, Ware LB: Elevated urinary IL-18 levels at the time of ICU admission predict adverse clinical outcomes. Clin J Am Soc Nephrol 2010;5:1497–1505.
45 Bennett M, Dent CL, Ma Q, Dastrala S, Grenier F, Workman R, Syed H, Ali S, Barasch J, Devarajan P: Urine NGAL predicts severity of acute kidney injury after cardiac surgery: a prospective study. Clin J Am Soc Nephrol 2008;3:665–673.
46 Kidney Disease: Improving Global Outcomes (KDIGO): http://www.renal.org/Clinical/GuidelinesSection/AcuteKidneyInjury.aspx (accessed September 3, 2011).
47 Sturgeon C, Hill R, Hortin GL, Thompson D: Taking a new biomarker into routine use – a perspective from the routine clinical biochemistry laboratory. Proteomics Clin Appl 2010;4:892–903.
48 Glasziou P, Irwig L, Deeks JJ: When should a new test become the current reference standard? Ann Intern Med 2008;149:816–822.
49 Goldstein SL, Chawla LS: Renal angina. Clin J Am Soc Nephrol 2010;5:943–949.
50 Acute Dialysis Quality Initiative (ADQI): www.ADQI.org (accessed January 10, 2013).

Peter A. McCullough, MD, MPH
Providence Park Heart Institute, Providence Park Hospital
47601 Grand River, B125
Novi, MI 48374 (USA)
E-Mail peteramccullough@gmail.com

McCullough PA, Kellum JA, Mehta RL, Murray PT, Ronco C (eds): ADQI Consensus on AKI Biomarkers and Cardiorenal Syndromes. Contrib Nephrol. Basel, Karger, 2013, vol 182, pp 30–44
DOI: 10.1159/000349964

Differential Diagnosis of AKI in Clinical Practice by Functional and Damage Biomarkers: Workgroup Statements from the Tenth Acute Dialysis Quality Initiative Consensus Conference

Zoltan H. Endre[a, b] · John A. Kellum[c] · Salvatore Di Somma[d] · Kent Doi[e] · Stuart L. Goldstein[f] · Jay L. Koyner[g] · Etienne Macedo[h] · Ravindra L. Mehta[i] · Patrick T. Murray[j] · for the ADQI 10 Workgroup

[a]Department of Nephrology, Prince of Wales Clinical School, University of New South Wales, Sydney, N.S.W., Australia; [b]Christchurch Kidney Research Group, Department of Medicine, University of Otago, Christchurch, New Zealand; [c]Department of Critical Care Medicine, School of Medicine, Pittsburgh, Pa., USA; [d]Department of Medical and Surgery Sciences and Translational Medicine, University La Sapienza, Rome, Italy; [e]Department of Emergency and Critical Care Medicine, The University of Tokyo, Tokyo, Japan; [f]Center for Acute Care Nephrology, Cincinnati Children's Hospital Medical Center, Cincinnati, Ohio, USA; [g]Section of Nephrology, Prizker Shool of Medicine, University of Chicago, Chicago, Ill., USA; [h]Department of Internal Medicine, Division of Nephrology, University of São Paulo School of Medicine, São Paulo, Brazil; [i]UCSD Medical Centre University of California San Diego, San Diego, Calif., USA; [j]School of Medicine and Medical Science, University College Dublin, Dublin, Ireland

Abstract

Acute kidney injury (AKI) is a common but complex clinical syndrome with multiple etiologies. These etiologies target different sites and pathways within the kidney. Novel biomarkers of 'kidney damage' (which can be tubular or glomerular) can be used to diagnose AKI, even in the absence of an increase in serum creatinine or oliguria. These biomarkers of kidney damage can be combined with biomarkers of kidney function to facilitate classification of AKI. A comprehensive review of the literature was performed using the published methodology of the Acute Dialysis Quality Initiative (ADQI) working group and used to establish consensus statements regarding the use of biomarkers in the differential diagnosis of AKI. We recommend that the pathophysiological terms 'functional change' and 'kidney damage' be used in preference to the anatomical classification using the terms pre-renal, renal and post-renal AKI. We further recommend the use of both renal and non-

Z.H.E. and J.A.K. are co-first authors; all other authors are listed in alphabetic order.
ADQI 10 Workgroup members are listed in Appendix 1.

renal biomarkers in establishing the specific cause of AKI as soon as possible after diagnosis. The presence of underlying CKD or of sepsis poses additional challenges in differential diagnosis, since these conditions alter both baseline biomarker excretion and biomarker performance. We recommend that biomarkers be validated within the clinical context in which they are to be used. Within that context, combinations of biomarkers may, in the future, allow differentiation of the site, mechanism and phase of injury.

Acute kidney injury (AKI) is a complex clinical syndrome that may arise in response to multiple exposures (e.g. sepsis, nephrotoxins, circulatory shock including acute cardiac failure) and whose underlying pathogenesis is incompletely understood [for recent reviews, see 1–3]. Sepsis is the major cause of AKI, accounting for nearly 50% of cases [4–6]. Approximately one third of hospitalized patients with community-acquired sepsis develop AKI [7, 8]. The various causes of AKI may affect kidney function differently and may injure the kidney in different ways. Biomarkers detect changes in kidney function (mainly glomerular filtration function, also tubular functions) and damage (mainly tubular, but also glomerular injury); both provide complementary information [9].

While clinical practice guidelines exist to aid in determining AKI etiology [10] these do not include kidney damage biomarkers. We propose a systematic approach that incorporates biomarkers not merely in the detection of AKI or risk of AKI, but also in establishing the specific cause or causes of AKI. We recognize that comorbidity, particularly the presence of underlying chronic kidney disease (CKD), poses an additional diagnostic challenge. This is the case, not only in terms of differentiating acute from chronic and from 'acute-on-chronic' disease, but also with respect to determining the etiology of AKI in the setting of underlying CKD. Again, we suggest that biomarkers may help, but special considerations apply.

Finally, we recognize that merely detecting AKI and defining its etiology does not address the nature of the pathophysiological state of the kidney at the time of diagnosis. This will vary in time and intensity with mechanism and duration of AKI, and must be determined in order to target treatment to the specific disease process affecting individual patients. Current biomarkers do not identify these and our proposals, suggestions and recommendations primarily address existing biomarkers. However, we will discuss the potential use of and need for new markers and will outline recommendations for future research.

Methods

Consistent with previous Acute Dialysis Quality Initiative (ADQI) meetings, a modified Delphi approach was followed. We performed a systematic search and review of the available literature pre-conference, as described in detail elsewhere in this volume. We

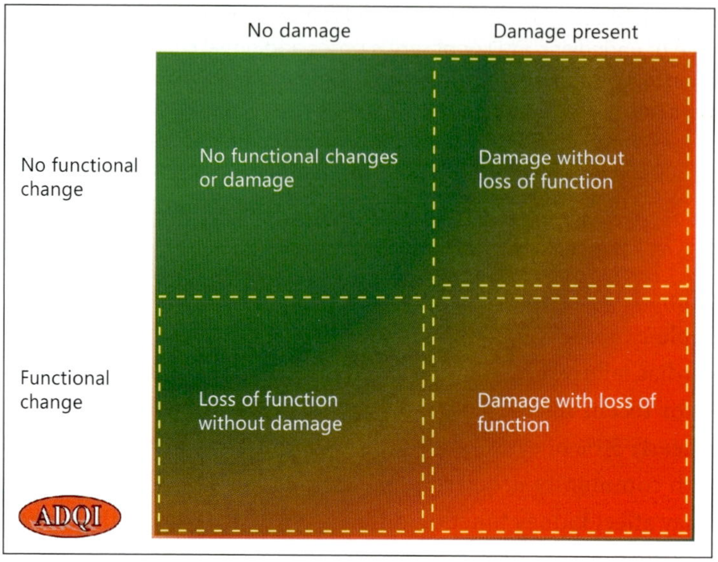

Fig. 1. As shown, the combination of functional and damage biomarkers allows categorization of patients into a specific category after the diagnosis of AKI is made. Currently this is based on diagnosis with standard creatinine and urine output criteria, however in future could include a diagnosis based on damage biomarker criteria. These findings would need to be combined with the clinical and laboratory information to determine the underlying pathophysiology and the contributory factors. Sequential assessments could provide information on which of the factors is prevalent for ongoing injury or resolution and offer opportunities for targeted intervention. Reproduced with permission from ADQI [59].

focused on the role of biomarkers in distinguishing etiology of AKI. Studies were identified via PubMed and Web of Science using the term 'biomarker', and acute kidney injury (AKI)' combined with 'etiology' and 'differential diagnosis'. The large body of literature retrieved generated a series of key questions, which were used to limit the scope of the review. Only representative publications are cited in this review.

Results

Our group started from the premise that AKI had already been diagnosed using a combination of biomarkers of acutely decreased kidney function and/or kidney damage (usually tubular), and defined using those terms (fig. 1). This left the following key questions for consideration with respect to differential diagnosis: (1) Can biomarkers distinguish a 'pre-renal state' from AKI? (2) Can biomarkers differentiate etiology of AKI? (3) Can biomarkers identify the mechanism and time course of AKI? (4) Can biomarkers distinguish AKI from CKD? (5) Can biomarkers distinguish de novo AKI from AKI superimposed on CKD? (6) What is role of biomarkers from other organs in the context of determining etiology of AKI?

Transient AKI, Not Pre-Renal

A systematic method for the differential diagnosis of AKI should optimize diagnostic assessment and therapeutic intervention. The term 'pre-renal' to classify both a group of causes of AKI leading to renal underperfusion and a unique state of reversible loss of kidney function with evidence of injury is both confusing and controversial despite a long tradition of usage [11, 12]. While grouping the causes of AKI as pre-renal, renal and post-renal causes can be helpful, its imprecision leads to limited diagnostic and therapeutic clarity. For instance, the treatment of intravascular dehydration, often cited as the classic example of pre-renal azotemia, includes rapid rehydration. However, other 'pre-renal' conditions, including nephrotic, hepatorenal and cardiorenal syndromes, can present with similar laboratory findings, yet the management for these conditions requires fluid restriction. Unfortunately, for many clinicians, the statement that a patient is 'pre-renal' has become conflated with being dehydrated.

Similarly, fluid responsiveness is not synonymous with 'pre-renal' AKI. In a recent prospective study of using urinary NGAL to distinguish pre-renal from intrinsic AKI, the investigators were able to classify the pre-renal versus intrinsic renal state based on fluid responsiveness in only 25% of cases [13]. Furthermore, the notion that aggressive fluid resuscitation leads to improved renal perfusion, reversal of the pre-renal state and prevention of intrinsic or established renal injury (or 'acute tubular necrosis' – ATN) has been refuted by animal data demonstrating that improved hemodynamics do not restore the renal microcirculation or renal blood flow after hypotensive shock [14, 15].

Reversibility of renal dysfunction does not define the absence of harm. Even if there is rapid recovery from AKI, there is a cost: several large cohort studies [16–19] and the prospective EARLYARF study [20] have identified that transient (<48 h) AKI was associated with the adverse outcomes of need for dialysis and death, even when AKI resolved within 24 h.

The concept of purely functional loss is also challenged by recent studies providing evidence that some, albeit less, damage is actually present in patients with transient AKI. If 'pre-renal' AKI is defined as transient AKI combined with evidence of preservation of renal tubular function (e.g. with a fractional sodium excretion less than 1%), then biomarkers of damage are increased above that observed in 'No-AKI' patients [20]. Even the earliest studies of reversible azotemia induced in volunteers by water deprivation noted kidney damage as shown by the development of hematuria and proteinuria, which reversed after rehydration [21]. On the other hand, there is good evidence that when AKI biomarkers are positive, even in the absence of apparent change in function, patients have worse hospital survival and even an increased need for dialysis than patients without AKI [22]. Of course, the biomarkers may simply predict changes in function that would manifest later except that competing endpoints such as death or dialysis occur first.

These arguments suggest that the concept of 'pre-renal AKI' is a flawed paradigm [12, 23], both in the presence and absence of sepsis, and that rapidly reversible AKI simply reflects a lesser degree of structural damage [20]. The observations highlight the need to move the approach to AKI differential diagnosis away from a semi-artificial anatomical construct and post hoc clinical determination to a proactive pathophysiological paradigm. An approach where biomarkers can be used to classify AKI by functional change, kidney damage, or both (fig. 1), provides information that is interpretable, and accounts for the dynamic nature of AKI. Both duration of AKI and degree of functional change and damage predict outcome.

We recommend that the terms 'functional change' and 'kidney damage' should be used in preference to the terms 'pre-, intra- and post-renal' in order to narrow the differential diagnosis of AKI. We encourage relevant further diagnostic assessment and therapeutic intervention as outlined below, without presupposing a static clinical state or waiting for a clinical response to treatment.

Etiology of AKI

It is critical that the etiology of AKI is determined as rapidly as possible, regardless of cause, since an important determinant of the response to therapy and long-term prognosis in many types of AKI is early diagnosis. For example, rapidly progressive glomerulonephritis due to anti-glomerular antibody disease can lead to irreversible renal failure within days, whereas immediate intervention with plasma exchange and immunosuppression preserves function [24]. Similarly, uncorrected outflow obstruction causing AKI leads eventually to renal parenchymal loss, while prompt treatment leads to rapid recovery [25]. Delayed diagnosis of AKI has contributed to the repeated failure of pharmaceutical intervention trials in AKI [9, 24]. In part, this delay results from diagnosis by exclusion which has been used in most intervention studies of AKI due to ischemia-reperfusion injury. Diagnosis by exclusion is an inadequate strategy for early diagnosis of any condition and AKI is no exception. While the benefit of early pharmaceutical intervention in AKI secondary to ischemia-reperfusion

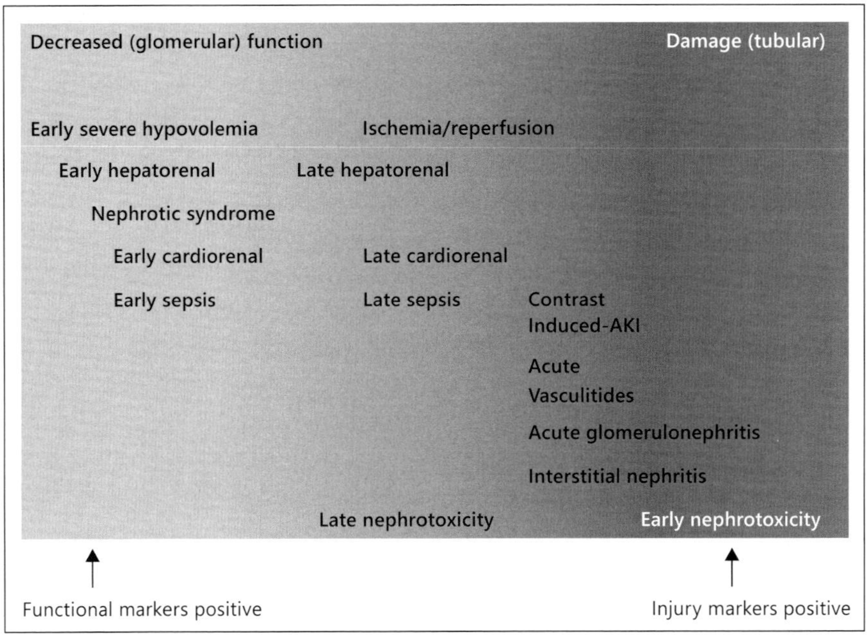

Fig. 2. Differential diagnosis of AKI. The arrows indicate the most likely type of biomarker to be positive during initial injury. Reproduced with permission from ADQI [59].

injury remains uncertain [26], early dialysis in established AKI may reduce mortality and accelerate recovery [27, 28].

Biomarkers offer the opportunity to diagnose AKI proactively in at risk patients (fig. 2). When the etiology of AKI affects glomerular function early, this will be reflected by early changes in biomarkers of filtration function, such as direct measurements of GFR and later by changes in creatinine or cystatin C. When tubular epithelium is damaged early, AKI will first be detected by changes in biomarkers of tubular injury, such as NGAL, KIM-1, IL-18, L-FABP, GGT and others [29]. In this setting, these biomarkers are non-specific markers of injury and context will identify the likely cause.

As an example, determining the etiology of renal failure in patients with cirrhosis, ascites, and edema remains an important clinical challenge, particularly in patients awaiting liver transplantation where glomerular disease is often present [30]. Differentiation of type 1 hepatorenal syndrome from other forms of AKI is potentially important since initial treatment of each is radically different [31]. A recent prospective cohort study of 115 adults with documented cirrhosis suggests that because the cut-offs were much higher, urinary NGAL could differentiate AKI from early heptorenal syndrome and also from pre-renal AKI and stable CKD in this setting [32].

In addition to facilitating early recognition, increased urinary or plasma injury biomarkers may identify involvement of specific nephron segments [33]. This may also help define the etiology of AKI. For example, in experimental AKI, nephrotoxins specific for different sites produce characteristic changing biomarker patterns of injury specific for each site [34]. A related approach is to utilize multiple biomarkers specific for different injury pathways to facilitate identify cause of AKI. A recent example is the use of increase in regenerating islet-derived protein III β (reg IIIβ) and gelsolin to differentiate between gentamicin and cisplatin nephrotoxicity [35]. Gentamicin increased the urinary concentrations of both reg IIIβ and gelsolin whereas these markers were not increased by cisplatin, nor were KIM-1 and NGAL increased by both nephrotoxins. reg IIIβ was found to be overexpressed in the kidneys of gentamicin-treated rats and secreted into the urine, whereas gelsolin is derived from plasma and appears in urine after glomerular filtration.

We recommend that the etiology of AKI be determined as soon as the diagnosis is made. Functional and injury biomarkers should be used to help differentiate AKI of uncertain etiology. Further studies will be required to determine the specificity of damage and biomarkers for individual disease states.

Setting, Mechanism and Time Course of AKI

Novel biomarkers of AKI have been validated in a limited variety of clinical settings, including cardiac surgery, sepsis and contrast-induced AKI. Several biomarkers have been shown to diagnose or predict the future development of AKI at early clinical timepoints (e.g. 2 h after initiation of cardiopulmonary bypass). However, not all biomarkers have performed well in every setting. Individual biomarkers have their own unique physiological and anatomical properties/fingerprints that determine their ability to differentiate the etiology of AKI. For instance, IL-18 is upregulated as part of the inflammatory response to cellular injury (activating macrophages, promoting T-cell differentiation and stimulating interferon-γ release by NK/T cells) [36]. Thus, IL-18 may not be well suited for diagnosing sepsis-associated AKI [37], in contrast to its performance in the setting of cardiac surgery-associated AKI where ischemia-reperfusion is presumed to be the dominant source of renal injury [37].

A summary of the current findings for the five most widely published biomarkers in several of the most widely investigated clinical settings is shown below (table 1). Note the inherent publication bias since negative biomarker findings often go unpublished. To date, there is no clinical evidence to support an individual biomarker or a panel of biomarkers in differentiating between dis-

Table 1. Novel AKI biomarker performance in different clinical settings

	NGAL	IL-18	KIM-1	CysC	L-FABP
Hypovolemic AKI	+/–	+/–	+/–	+/–	+/–
Sepsis	+	+/–	?	+	+
Post-cardiopulmonary bypass	+	+	+	+	+
Contrast-induced	+	+	+	+	+
Nephrotoxic	+	+	+	?	+
Delayed renal graft function	+	+	+	+	?

The table summarizes the literature as related to the most widely published biomarkers in extensively studied clinical settings. Cardiac surgery-associated AKI (especially after cardiopulmonary bypass) is the most widely published setting, and NGAL the most widely published biomarker. The table makes no attempt to compare the data for individual settings or biomarkers. Since the time course of each biomarker varies, the optimum sampling time will vary for each clinical context. NGAL and cystatin C data refer to both urinary and plasma values, the other biomarkers represent urinary values.
NGAL = Neutrophil gelatinase-associated lipocalin; IL-18 = interleukin-18; KIM-1 = kidney injury molecule 1; CysC = cystatin C; L-FABP = liver fatty acid-binding protein. + = Able to detect AKI in this clinical setting; – = not able to detect AKI in this clinical setting; ? = inconclusive data on ability to detect AKI in this setting.

Table 2. Five urinary biomarkers in AKI complicated by sepsis

	Non-sepsis No-AKI (n = 178)	Non-sepsis AKI (n = 95)	Sepsis No-AKI (n = 30)	Sepsis AKI (n = 36)
L-FABP, ng/ml	6.8 (1.5–17.3)	35.9 (6.8–156.0)[b]	21.5 (4.6–46.1)	64.5 (18.8–543.3)[a]
NGAL, ng/ml	11.5 (4.7–33.2)	32.8 (13.3–136.9)	137.8 (39.2–323.5)	271.0 (118.1–1,593.6)[a]
IL-18, pg/ml	45.3 (20.0–110.6)	122.9 (34.5–390.2)	206.2 (133.5–802.1)	405.0 (209.6–1,043.9)
NAG, U/l	7.5 (3.8–14.7)	11.9 (5.4–22.6)[b]	22.8 (9.3–40.2)	22.4 (11.9–45.8)
Albumin, mg/dl	2.9 (1.1–7.1)	6.2 (3.0–26.5)[b]	7.8 (3.5–21.1)	15.4 (7.4–43.7)

Values are presented as median (interquartile range). [a] $p < 0.05$ vs. non-sepsis/non-AKI, non-sepsis/AKI, and sepsis/non-AKI. [b] $p < 0.05$ vs. non-sepsis/non-AKI. Data from Doi et al. [48].

tinct clinical etiologies of AKI (e.g. cardiac surgery-associated versus radiocontrast). Larger studies with higher sensitivity biomarker assays are needed to validate current findings and to define the role of these biomarkers in the differential diagnosis of AKI. Note also that the threshold concentrations for AKI diagnosis with these biomarkers are likely to be context-specific, particularly in the context of sepsis where baseline biomarker concentrations are higher in the No-AKI group (see table 2 and the next section).

Currently available biomarkers are used as non-specific markers of kidney damage. Thus, although each of the most widely studied biomarkers (especially those in table 1) has at least one defined role in the pathway leading to or from injury, these biomarkers are presently used for diagnosis or prediction of AKI or outcome and not to identify the specific pathways involved. Biomarker performance is known to vary, inter alia, with baseline renal function, duration of injury, and etiology [38]. Each biomarker has a specific time course in a specific setting of injury [38–40]. From a consideration of biomarkers in nephrotoxic AKI, it is likely that these biomarker profiles will be setting-dependent [34, 41, 42]. We recommend using *only* biomarkers with validated test characteristics for detection of AKI in the clinical context where they are being used (e.g. sepsis, contrast) and indeed the best validation strategy will include unselected patients with multiple etiologies since these are the varying types of patients that will be encountered in clinical practice.

Research recommendation: we recommend that biomarkers be validated in multiple different etiologies.

Differential Diagnosis in the Presence of CKD

The etiology of AKI is needed to identify the optimal therapeutic strategies for patient care. Several clinical studies have reported biomarker levels for different AKI etiologies. Most were dedicated to differentiating supposedly functional pre-renal AKI from established AKI. Urinary NGAL in the emergency department was significantly higher in patients with established AKI compared with pre-renal azotemia, regardless of cause [13, 43]. These two urinary NGAL studies showed that a urinary NGAL level >104 ng/ml supported established AKI, whereas a urinary NGAL level <47 ng/ml was suggestive of other conditions. Nevertheless, about 25% of AKI patients in the latter cohort were clinically unclassifiable and had urinary NGAL levels in an intermediate range.

Because AKI occurring in the ICU is frequently complicated by sepsis [4] and since sepsis and AKI synergistically increase mortality rate [44], identifying sepsis as the cause of AKI is of great potential significance in clinical management. Plasma NGAL has been reported to be higher in septic than in non-septic AKI patients in several [44, 45], but not all [46] studies. Both sepsis and AKI increased urinary cystatin C and their effects appear to be additive [47]. Similarly, evaluation of five urinary biomarkers (NGAL, IL-18, L-FABP, NAG, and albumin) in a mixed ICU revealed that the highest values were all in septic AKI patients, with L-FABP able to discriminate AKI in both septic and non-septic patients (table 1) [48]. These results suggest that higher thresholds of single and multiple biomarkers may differentiate septic from non-septic AKI.

Observational studies have described CKD in approximately 30% of the AKI patients in the ICU [4, 49]. Both baseline concentration and injury response vary amongst biomarkers in the presence of reduced renal function. Baseline urinary and plasma NGAL levels are known to be increased in CKD patients under stable conditions [50–52]. Urinary L-FABP is higher in CKD than in healthy control subjects [52]. Injury response and biomarker performance are modified in the presence of CKD (baseline eGFR <60 ml/min) [38, 53, 54]. A probable additional cause of increased baseline urinary biomarker concentrations in CKD is the presence of proteinuria, a common feature in CKD. Urinary albumin and protein competitively increase the concentration of filtered low molecular weight protein biomarkers such as NGAL and cystatin C, proteins normally reabsorbed by megalin-cubulin transport in the proximal tubule [55]. Thus in CKD patients, higher cut-off values are probably necessary to detect AKI superimposed on CKD as distinct from de novo AKI. Medication may also modify baseline biomarker concentration in the case of plasma NGAL (but not cystatin C); a reduction in plasma NGAL was observed following chronic atorvastatin treatment [54]. Thus both negative and positive associations with baseline renal function are possible. Negative associations are more common though biomarker concentration may increase after an initial delay in in patients with CKD [38].

Taken together, these features indicate that the magnitude and range of biomarker increase in AKI is modified by reduced GFR in CKD. It is therefore essential that the pattern of biomarker increase is evaluated over time when the CKD status is unknown. The slope of the increase, peak value, time to peak, as well as slope of decrease and time to decrease may help differentiate between AKI superimposed on CKD and de novo AKI. Thus, the biomarker threshold and pattern over time as well as the baseline level must be considered when determining the etiology of AKI in patients with underlying CKD.

We recommend that research studies should be performed to determine the thresholds, biomarker profiles and disease profiles that differentiate among etiologies. We also recommend studies to clarify the frequency of measurement of biomarkers and that baseline levels of biomarkers be established for patients with various stages of CKD. Biomarkers need to be validated in cohorts with CKD.

Non-Renal Biomarkers and Etiology of AKI

In the differential diagnosis of AKI, context-specific biomarkers will assist in determining AKI etiology. Similarly in the context of a specific clinical presentation, structural and functional biomarkers of renal injury will define the presence of complicating AKI (fig. 3). The diagnostic and prognostic roles of natri-

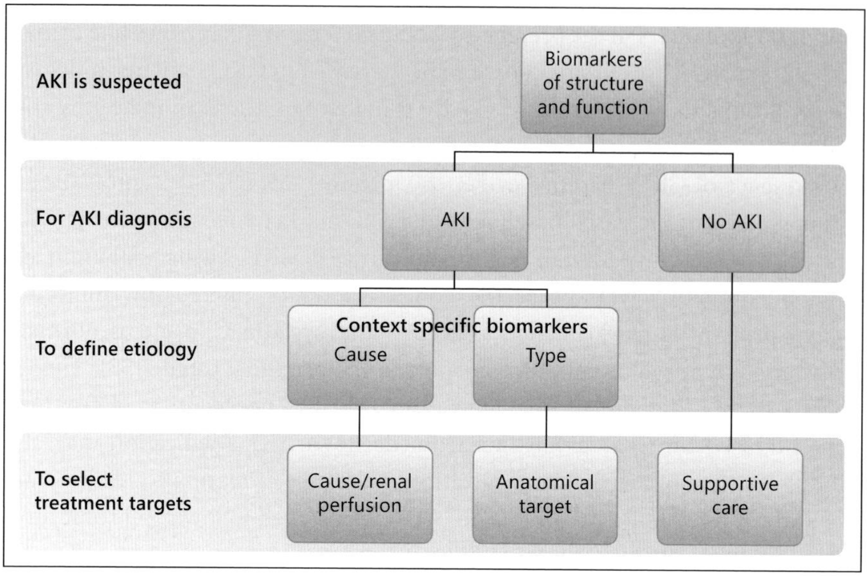

Fig. 3. Biomarker utility in differential diagnosis of established AKI and in the identification of injury pathways. Reproduced with permission from ADQI [59].

uretic peptides (BNP, NT-proBNP, MD-proANP) and procalcitonin (PCT) as biomarkers are well established in patients with heart failure and sepsis [56–58]. Renal injury is common in these patients and will result in poorer outcomes. Combining biomarkers of cardiac failure or sepsis with those of structural and functional kidney damage should facilitate both early detection of AKI and facilitate differential diagnosis.

Both acute or chronic cardiac failure compromise renal perfusion with consequent sodium and water retention, arterial vasoconstriction, venous congestion and impaired kidney function defining cardiorenal syndromes type I and II. There is an inverse relationship between renal function and NT-proBNP values in cardiac failure patients, where BNP has a prognostic role and predicts the development of AKI [53]. The combination of BNP with NGAL has a diagnostic and prognostic utility in patients with cardiorenal syndrome [54].

Conclusion

We propose a systematic approach to the diagnosis of AKI that incorporates a combination of functional and damage biomarkers for the diagnosis of AKI and in establishing the specific cause or causes of AKI. The presence of underlying CKD poses an additional diagnostic challenge, not only in terms of differentiating acute from chronic and from 'acute on chronic' disease, but also with respect to determining the etiology of AKI in the setting of underlying CKD. We recommend that functional and damage biomarkers be validated within the context or domain in which they are to be used. Within that context, combinations of biomarkers may, in the future, allow differentiation of the site, mechanism and timing of injury.

References

1 Murugan R, Kellum JA: Acute kidney injury: what's the prognosis? Nat Rev Nephrol 2011; 7:209–217.
2 Singbartl K, Kellum JA: AKI in the ICU: definition, epidemiology, risk stratification, and outcomes. Kidney Int 2012;81:819–825.
3 Bonventre JV, Yang L: Cellular pathophysiology of ischemic acute kidney injury. J Clin Invest 2011;121:4210–4221.
4 Uchino S, Kellum JA, Bellomo R, Doig GS, Morimatsu H, Morgera S, Schetz M, Tan I, Bouman C, Macedo E, Gibney N, Tolwani A, Ronco C, Beginning and Ending Supportive Therapy for the Kidney BEST Kidney Investigators: Acute renal failure in critically ill patients: a multinational, multicenter study. JAMA 2005;294:813–818.
5 Ali T, Khan I, Simpson W, Prescott G, Townend J, Smith W, Macleod A: Incidence and outcomes in acute kidney injury: a comprehensive population-based study. J Am Soc Nephrol 2007;18:1292–1298.
6 Bagshaw SM, Uchino S, Bellomo R, Morimatsu H, Morgera S, Schetz M, Tan I, Bouman C, Macedo E, Gibney N, Tolwani A, Oudemans-van Straaten HM, Ronco C, Kellum JA, Beginning and Ending Supportive Therapy for the Kidney BEST Kidney Investigators: Septic acute kidney injury in critically ill patients: clinical characteristics and outcomes. Clin J Am Soc Nephrol 2007; 2:431–439.
7 Murugan R, Karajala-Subramanyam V, Lee M, Yende S, Kong L, Carter M, Angus DC, Kellum JA, Investigators OBOTGAIMOSG: Acute kidney injury in non-severe pneumonia is associated with an increased immune response and lower survival. Kidney Int 2009;77:527–535.
8 Lopes JA, Jorge S, Resina C, Santos C, Pereira A, Neves J, Antunes F, Prata MM: Prognostic utility of RIFLE for acute renal failure in patients with sepsis. Crit Care 2007;11:408.
9 Endre ZH, Pickering JW, Walker RJ: Clearance and beyond: the complementary roles of GFR measurement and injury biomarkers in acute kidney injury. Am J Physiol Renal Physiol 2011;301:F697–F707.
10 KDIGO Section 2: AKI definition. Kidney Int Suppl 2012;2:7–24.
11 Parikh CR, Coca SG: Acute kidney injury: defining prerenal azotemia in clinical practice and research. Nat Rev Nephrol 2010;6:641–642.
12 Bellomo R, Bagshaw S, Langenberg C, Ronco C: Pre-renal azotemia: a flawed paradigm in critically ill septic patients? Contrib Nephrol. Basel, Karger, 2007, vol 156, pp 1–9.
13 Singer E, Elger A, Elitok S, Kettritz R, Nickolas TL, Barasch J, Luft FC, Schmidt-Ott KM: Urinary neutrophil gelatinase-associated lipocalin distinguishes pre-renal from intrinsic renal failure and predicts outcomes. Kidney Int 2011;80:405–414.
14 Legrand M, Mik EG, Balestra GM, Lutter R, Pirracchio R, Payen D, Ince C: Fluid resuscitation does not improve renal oxygenation during hemorrhagic shock in rats. Anesthesiology 2010;112:119.
15 Wan L, Bellomo R, May CN: The effect of normal saline resuscitation on vital organ blood flow in septic sheep. Intensive Care Med 2006;32:1238–1242.
16 Uchino S, Bellomo R, Bagshaw SM, Goldsmith D: Transient azotaemia is associated with a high risk of death in hospitalized patients. Nephrol Dial Transplant 2010;25:1833–1839.

17 Brown JR, Kramer RS, Mackenzie TA, Coca SG, Sint K, Parikh CR: Determinants of acute kidney injury duration after cardiac surgery: an externally validated tool. Ann Thorac Surg 2012;93:570–576.
18 Brown JR, Kramer RS, Coca SG, Parikh CR: Duration of acute kidney injury impacts long-term survival after cardiac surgery. Ann Thorac Surg 2010;90:1142–1148.
19 Coca SG, King JT, Rosenthal RA, Perkal MF, Parikh CR: The duration of postoperative acute kidney injury is an additional parameter predicting long-term survival in diabetic veterans. Kidney Int 2010;78:926–933.
20 Nejat M, Pickering JW, Devarajan P, Bonventre JV, Edelstein CL, Walker RJ, Endre ZH: Some biomarkers of acute kidney injury are increased in pre-renal acute injury. Kidney Int 2012;81:1254–1262.
21 Coller FA, Maddock WG: A study of dehydration in humans. Ann Surg 1935;102: 947–960.
22 Haase M, Devarajan P, Haase-Fielitz A, Bellomo R, Cruz DN, Wagener G, Krawczeski CD, Koyner JL, Murray P, Zappitelli M, Goldstein SL, Makris K, Ronco C, Martensson J, Martling CR, Venge P, Siew E, Ware LB, Ikizler TA, Mertens PR: The outcome of neutrophil gelatinase-associated lipocalin-positive subclinical acute kidney injury: a multicenter pooled analysis of prospective studies. J Am Coll Cardiol 2011; 57:1752–1761.
23 Kellum JA: Prerenal azotemia: still a useful concept? Crit Care Med 2007;35:1630–1631.
24 Pusey C: Anti-glomerular basement membrane disease. Kidney Int 2003;64:1535–1550.
25 Sacks SH, Aparicio SA, Bevan A, Oliver DO, Will EJ, Davison AM: Late renal failure due to prostatic outflow obstruction: a preventable disease. BMJ 1989;298:156–159.
26 Endre ZH, Walker RJ, Pickering JW, Shaw GM, Frampton CM, Henderson SJ, Hutchison R, Mehrtens JE, Robinson JM, Schollum JBW, Westhuyzen J, Celi LA, McGinley RJ, Campbell IJ, George PM: Early intervention with erythropoietin does not affect the outcome of acute kidney injury (the EARLYARF trial). Kidney Int 2010;77:1020–1030.
27 Bagshaw SM, Uchino S, Bellomo R, Morimatsu H, Morgera S, Schetz M, Tan I, Bouman C, Macedo E, Gibney N: Timing of renal replacement therapy and clinical outcomes in critically ill patients with severe acute kidney injury. J Crit Care 2009;24:129–140.
28 Seabra VF, Balk EM, Liangos O, Sosa MA, Cendoroglo M, Jaber BL: Timing of renal replacement therapy initiation in acute renal failure: a meta-analysis. Am J Kidney Dis 2008;52:272–284.
29 Coca S, Yalavarthy R, Concato J, Parikh CR: Biomarkers for the diagnosis and risk stratification of acute kidney injury: a systematic review. Kidney Int 2008;73:1008–1016.
30 Crawford DH, Endre ZH, Axelsen RA, Lynch SV, Balderson GA, Strong RW, Kerlin P, Powell LW, Fleming SJ: Universal occurrence of glomerular abnormalities in patients receiving liver transplants. Am J Kidney Dis 1992;19:339–344.
31 Guevara M, Arroyo V: Hepatorenal syndrome. Expert Opin Pharmacother 2011; 12:1405–1417.
32 Verna EC, Brown RS, Farrand E, Pichardo EM, Forster CS, Sola-Del Valle DA, Adkins SH, Sise ME, Oliver JA, Radhakrishnan J, Barasch JM, Nickolas TL: Urinary neutrophil gelatinase-associated lipocalin predicts mortality and identifies acute kidney injury in cirrhosis. Dig Dis Sci 2012;57:2362–2370.
33 Bonventre JV, Vaidya VS, Schmouder R, Feig P, Dieterle F: Next-generation biomarkers for detecting kidney toxicity. Nat Biotechnol 2010;28:436–440.
34 Sasaki D, Yamada A, Umeno H, Kurihara H, Nakatsuji S, Fujihira S, Tsubota K, Ono M, Moriguchi A, Watanabe K, Seki J: Comparison of the course of biomarker changes and kidney injury in a rat model of drug-induced acute kidney injury. Biomarkers 2011;16:553–566.
35 Ferreira L, Quiros Y, Sancho-Martínez SM, García-Sánchez O, Raposo C, López-Novoa JM, González-Buitrago JM, Lopez-Hernandez FJ: Urinary levels of regenerating islet-derived protein III β and gelsolin differentiate gentamicin from cisplatin-induced acute kidney injury in rats. Kidney Int 2011; 79:518–528.

36 Martinon F, Mayor A, Tschopp J: The inflammasomes: guardians of the body. Annu Rev Immunol 2009;27:229–265.

37 Siew E, Ikizler TA, Gebretsadik T, Shintani A, Wickersham N, Bossert F, Peterson JF, Parikh CR, May AK, Ware LB: Elevated urinary IL-18 levels at the time of ICU admission predict adverse clinical outcomes. Clin J Am Soc Nephrol 2010;5:1497–1505.

38 Endre ZH, Pickering JW, Walker RJ, Devarajan P, Edelstein CL, Bonventre JV, Frampton CM, Bennett MR, Ma Q, Sabbisetti VS, Vaidya VS, Walcher AM, Shaw GM, Henderson SJ, Nejat M, Schollum JBW, George PM: Improved performance of urinary biomarkers of acute kidney injury in the critically ill by stratification for injury duration and baseline renal function. Kidney Int 2011;79:1119–1130.

39 Krawczeski CD, Goldstein SL, Woo JG, Wang Y, Piyaphanee N, Ma Q, Bennett M, Devarajan P: Temporal relationship and predictive value of urinary acute kidney injury biomarkers after pediatric cardiopulmonary bypass. J Am Coll Cardiol 2011;58:2301–2309.

40 McIlroy DR, Wagener G, Lee HT: Biomarkers of acute kidney injury: an evolving domain. Anesthesiology 2010;112:998–1004.

41 Ozer JS, Dieterle F, Troth S, Perentes E, Cordier A, Verdes P, Staedtler F, Mahl A, Grenet O, Roth DR, Wahl D, Legay F, Holder D, Erdos Z, Vlasakova K, Jin H, Yu Y, Muniappa N, Forest T, Clouse HK, Reynolds S, Bailey WJ, Thudium DT, Topper MJ, Skopek TR, Sina JF, Glaab WE, Vonderscher J, Maurer G, Chibout SD, Sistare FD, Gerhold DL: A panel of urinary biomarkers to monitor reversibility of renal injury and a serum marker with improved potential to assess renal function. Nat Biotechnol 2010;28:486–494.

42 Hoffmann D, Fuchs TC, Henzler T, Matheis KA, Herget T, Dekant W, Hewitt P, Mally A: Evaluation of a urinary kidney biomarker panel in rat models of acute and subchronic nephrotoxicity. Toxicology 2010;277:49–58.

43 Nickolas TL, Schmidt-Ott KM, Canetta P, Forster C, Singer E, Sise M, Elger A, Maarouf O, Sola-Del Valle DA, O'Rourke M, Sherman E, Lee P, Geara A, Imus P, Guddati A, Polland A, Rahman W, Elitok S, Malik N, Giglio J, El-Sayegh S, Devarajan P, Hebbar S, Saggi SJ, Hahn B, Kettritz R, Luft FC, Barasch J: Diagnostic and prognostic stratification in the emergency department using urinary biomarkers of nephron damage: a multicenter prospective cohort study. J Am Coll Cardiol 2012;59:246–255.

44 Bagshaw SM, Bennett M, Haase M, Haase-Fielitz A, Egi M, Morimatsu H, D'Amico G, Goldsmith D, Devarajan P, Bellomo R: Plasma and urine neutrophil gelatinase-associated lipocalin in septic versus non-septic acute kidney injury in critical illness. Intensive Care Medicine 2010;36:452–461.

45 Kümpers P, Hafer C, Lukasz A, Lichtinghagen R, Brand K, Fliser D, Faulhaber-Walter R, Kielstein JT: Serum neutrophil gelatinase-associated lipocalin at inception of renal replacement therapy predicts survival in critically ill patients with acute kidney injury. Crit Care 2010;14:R9.

46 Martensson J, Bell M, Oldner A, Xu S, Venge P, Martling CR: Neutrophil gelatinase-associated lipocalin in adult septic patients with and without acute kidney injury. Intensive Care Med 2010;36:1333–1340.

47 Nejat M, Pickering JW, Walker RJ, Westhuyzen J, Shaw GM, Frampton CM, Endre ZH: Urinary cystatin C is diagnostic of acute kidney injury and sepsis, and predicts mortality in the intensive care unit. Crit Care 2010;14:R85.

48 Doi K, Negishi K, Ishizu T, Katagiri D, Fujita T, Matsubara T, Yahagi N, Sugaya T, Noiri E: Evaluation of new acute kidney injury biomarkers in a mixed intensive care unit. Crit Care Med 2011;39:2464–2469.

49 Mehta RL, Pascual MT, Soroko S, Savage BR, Himmelfarb J, Ikizler TA, Paganini EP, Chertow GM, Program to Improve Care in Acute Renal Disease: Spectrum of acute renal failure in the intensive care unit: the PICARD experience. Kidney Int 2004;66:1613–1621.

50 Bolignano D, Lacquaniti A, Coppolino G, Donato V, Campo S, Fazio MR, Nicocia G, Buemi M: Neutrophil gelatinase-associated lipocalin and progression of chronic kidney disease. Clin J Am Soc Nephrol 2009;4:337–344.

51 Malyszko J, Bachorzewska-Gajewska H, Sitniewska E, Malyszko JS, Poniatowski B, Dobrzycki S: Serum neutrophil gelatinase-associated lipocalin as a marker of renal function in non-diabetic patients with stage 2–4 chronic kidney disease. Ren Fail 2008;30: 625–628.
52 Kamijo A, Sugaya T, Hikawa A, Yamanouchi M, Hirata K, Ishimitsu T, Numabe A, Takagi M, Hayakawa H, Tabei F: Clinical evaluation of urinary excretion of liver-type fatty acid-binding protein as a marker for the monitoring of chronic kidney disease: a multicenter trial. J Lab Clin Med 2005;145: 125–133.
53 McIlroy DR, Wagener G, Lee HT: Neutrophil gelatinase-associated lipocalin and acute kidney injury after cardiac surgery: the effect of baseline renal function on diagnostic performance. Clin J Am Soc Nephrol 2010;5: 211–219.
54 Fassett RG, Venuthurupalli SK, Gobé GC, Coombes JS, Cooper MA, Hoy WE: Biomarkers in chronic kidney disease: a review. Kidney Int 2011;80:806–821.
55 Nejat M, Hill JV, Pickering JW, Edelstein CL, Devarajan P, Endre ZH: Albuminuria increases cystatin C excretion: implications for urinary biomarkers. Nephrol Dial Transpl 2011.
56 Reinhart K, Meisner M: Biomarkers in the critically ill patient: procalcitonin. Crit Care Clin 2011;27:253–263.
57 Maisel AS, Mueller C, Fitzgerald R, Brikhan R, Hiestand BC, Iqbal N, Clopton P, van Veldhuisen DJ: Prognostic utility of plasma neutrophil gelatinase-associated lipocalin in patients with acute heart failure: the NGAL EvaLuation Along with B-type NaTriuretic Peptide in acutely decompensated heart failure (GALLANT) trial. Eur J Heart Fail 2011;13:846–851.
58 Maisel A, Mueller C, Adams K, Anker SD, Aspromonte N, Cleland JGF, Cohen-Solal A, Dahlstrom U, DeMaria A, Di Somma S, Filippatos GS, Fonarow GC, Jourdain P, Komajda M, Liu PP, McDonagh T, McDonald K, Mebazaa A, Nieminen MS, Peacock WF, Tubaro M, Valle R, Vanderhyden M, Yancy CW, Zannad F, Braunwald E: State of the art: using natriuretic peptide levels in clinical practice. Eur J Heart Fail 2008;10:824–839.
59 Acute Dialysis Quality Initiative (ADQI): www.ADQI.org (accessed January 10, 2013).

Zoltán H. Endre
Department of Nephrology, Prince of Wales Hospital
High Street, Randwick
Sydney, NSW 2031 (Australia)
E-Mail z.endre@unsw.edu.au

McCullough PA, Kellum JA, Mehta RL, Murray PT, Ronco C (eds): ADQI Consensus on AKI Biomarkers and Cardiorenal Syndromes. Contrib Nephrol. Basel, Karger, 2013, vol 182, pp 45–64
DOI: 10.1159/000349965

Use of Biomarkers to Assess Prognosis and Guide Management of Patients with Acute Kidney Injury

Dinna N. Cruz[a, b] · Sean M. Bagshaw[c] · Alan Maisel[d] · Andrew Lewington[e] · Ravi Thadhani[f] · Rajasekara Chakravarthi[g] · Patrick T. Murray[h] · Ravindra L. Mehta[i] · Lakhmir S. Chawla[j] · for the ADQI 10 Workgroup

[a]Department of Nephrology Dialysis & Transplantation, San Bortolo Hospital, Vicenza, Italy; [b]International Renal Research Institute (IRRIV), Vicenza, Italy; [c]Division of Critical Care Medicine, Faculty of Medicine and Dentistry, University of Alberta, Edmonton, Alta., Canada; [d]Department of Medicine and Cardiology, San Diego VA Medical Center and University of California, San Diego, Calif., USA; [e]Department of Renal Medicine, St James's University Hospital, Leeds, UK; [f]Division of Nephrology, Massachusetts General Hospital and Harvard Medical School (RT), Boston, Mass., USA; [g]Department of Nephrology, CARE Hospital, The Institute of Medical Sciences, Hyderabad, India; [h]School of Medicine & Medical Science, University College Dublin, Dublin, Ireland; [i]Department of Medicine and Nephrology, University of California, San Diego, Calif., USA; [j]Department of Anesthesiology and Critical Care Medicine, George Washington University Medical Center, Washington, D.C., USA

Abstract

Several new biomarkers of kidney damage have been characterized and are being validated in clinical studies. These damage biomarkers complement existing conventional biomarkers of kidney function (e.g. serum creatinine, serum urea, and urine output) that are currently utilized to diagnose and stage acute kidney injury (AKI). Both functional and damage biomarkers provide an opportunity to identify patients with AKI who are at risk for a less favorable prognosis in terms of worsening damage or further declines in kidney function and likelihood of need for renal replacement. We performed a systemic search and review of the available literature pre-conference. Our workgroup presented the findings in multiple rounds to the ADQI conference members and a final summary and review was refined in an iterative approach. The specific clinical situations of renal or liver transplantation, or cirrhosis/hepatorenal syndrome were not included. Overall, multiple AKI biomarkers have been well characterized for utilization for AKI prognosis. These function-

D.N.C. and L.S.C. were workgroup facilitators.
ADQI 10 Workgroup members are listed in Appendix 1.

al and damage markers can be used to assist in decisions related to triage of patients with AKI and identifying patients with who are at risk for progression. Set cut-offs for various biomarkers and their bedside utility are forthcoming and will be in part determined by regulatory intended use guidelines, platform standardization, and inter-laboratory calibration. There remain many unresolved areas of AKI biomarker use in selected syndromes of AKI (e.g. cardiorenal syndrome, hepatorenal syndrome). As clinicians gain experience with AKI biomarkers, clinical care plans that incorporate them into routine care will shortly follow.

The current annual expense in the USA for the treatment of acute kidney injury (AKI) is estimated at USD 10 billion [1]. Patients who develop AKI have worse outcomes characterized by higher mortality, prolonged duration of hospitalization and increased risk for progression to chronic kidney disease (CKD) [2, 3]. Despite advances in modern medicine the outcomes from AKI have remained relatively unchanged over the past few decades [4]. The failure to translate successful interventional strategies from animal models to humans has in part been attributed to the lack of appropriately sensitive and specific biomarkers for the timely identification of kidney damage. Over the last decade, several new biomarkers of kidney damage have been characterized and are being validated in clinical studies. These damage biomarkers complement existing conventional biomarkers of kidney function (e.g. serum creatinine, serum urea, and urine output) that are currently utilized to diagnose and stage AKI. Both functional and damage biomarkers provide an opportunity to identify patients with AKI who are at risk for a less favorable prognosis in terms of worsening damage or further declines in kidney function, likelihood of need for renal replacement therapy (RRT), non-recovery of kidney function, greater resource utilization, and mortality. However, there are currently no evidence-based or consensus-driven recommendations for the pragmatic utilization of these biomarkers for managing patients with AKI such as identifying time points for interventions predicting outcomes. As additional kidney damage-specific biomarkers are identified, the integration of both functional and novel damage biomarkers will provide novel opportunities to improve the risk identification, the risk stratification and management, along with better prognosis information for patients with AKI. Biomarkers could permit the risk stratification of patients at an earlier stage for referral to renal services and avoid unnecessary delays in treatment and patient transfer. Conversely, the ability of a biomarker to determine that recovery is imminent could forestall the inappropriate commencement of RRT and the risks associated with line insertion (infection and bleeding). Alternatively, the biomarker could indicate that recovery is unlikely in the short term and therefore commencement of RRT is in the best interests of the patient before the development of uremic complications and exposure to non-beneficial medical management such as excessive fluid administration and/or diuretics. Furthermore, the severity of AKI predicts progression to CKD [5] and thus there is a clinical need to identify patients who are at a higher risk of developing progressive kidney disease, rather than complete recovery from AKI. Biomarkers could help identify risk for incomplete recovery and long-term progression to CKD and allow the earlier introduction of conventional or even novel interventional therapy.

In this article, we focus on defining the role of functional and damage biomarkers in predicting the course and consequences of AKI. Based on currently available evidence, we propose combining functional and damage markers to characterize the course of patients with AKI and guide timely interventions. We recognize that there are several gaps in our knowledge and we provide suggestions for further research.

Methods

We performed a systemic search and review of the available literature pre-conference, as described elsewhere in this volume. Specifically, we used the terms 'biomarkers', 'acute kidney injury', and 'acute renal failure', combined with 'prognosis', 'renal replacement therapy', 'dialysis', 'death' and 'mortality'. In view of the volume of the retrieved literature, only representative publications are cited in this review. Furthermore, we opted to limit this summary to clinical scenarios in which there was ample data. We did not include the specific clinical situations of renal or liver transplantation, or cirrhosis/hepatorenal syndrome.

Results

Based on the literature identified prior to the conference, the following key questions were considered: (1) When the diagnosis of AKI is established, are biomarkers able to reliably predict the occurrence of clinically relevant patient-centered outcomes (including worsening AKI, RRT utilization, kidney non-recovery (CKD), mortality and/or a composite of these outcomes)? (2) Which biomarkers (and at what thresholds) can be used by clinicians to predict these patient-centered outcomes? Is there sufficient evidence to support context-specific absolute cut-off and/or threshold values for prognostication in AKI? (3) Are there clinical factors which confound the operative characteristics of these biomarkers for the diagnosis of AKI? (4) Do repeated biomarker measurements provide added and/or incremental prognostic information for AKI? (5) What clinical actions, if any, can the clinician take based on biomarker measurements? Can biomarkers aid in the triage of at-risk patients and/or inform clinical management decisions?

Table 1. Selected references on biomarkers for the prediction of clinical outcomes

	RRT	Death	Worsening AKI and/or RRT and/or death	Nephrology consult, ICU admission, RRT or death	Renal recovery	Length of stay and/or mechanical ventilation
Urine output		16				
Urine microscopy			13, 14			
NGAL	17, 18, 38, 48	10, 17, 18, 48, 57	7, 10, 17, 38, 48	51	26	6, 7, 29
IL-18	10	36, 40, 57	8, 10, 11, 40			6, 29
KIM-1			8			
Cystatin C	10, 21		9, 10, 22, 23		26	
L-FABP		57				
NAG				51		
HGF					26	
π-GST			9			

Prediction of Clinically Relevant Outcomes

While there is an abundance of publications on novel kidney injury biomarkers, most of these data have focused on their ability to predict the development of AKI, generally defined by changes in the currently used functional biomarkers of kidney function (serum creatinine and/or urine output). Some data, however, do exist on the ability of selected biomarkers to provide information above and beyond simple prediction of AKI. These data provide additional prognostic information, including the severity and duration of AKI, likelihood of need to initiation RRT, delayed and/or non-recovery of kidney function and mortality (table 1). In addition, conventional biomarkers of kidney function (i.e. change in serum creatinine, urine output) have been reassessed to determine their predictive capacity in a contemporary format.

Many of these so-called 'prognostic' biomarker studies have been performed in the setting of timed injury models, such as after cardiac surgery where the onset of injury stimulus is known, and in the ICU settings where multiple insults combine to lead to AKI. For example, in the setting of pediatric cardiac surgery, early postoperative elevations in urinary interleukin (IL)-18, and urinary and plasma neutrophil gelatinase-associated lipocalin (NGAL) were all significantly associated with longer lengths of ICU stay after adjustment for other prognostic factors, and quintiles of each biomarker were also associated with duration of mechanical ventilation postoperatively in multivariable analyses [6]. Also in the setting of pediatric cardiac surgery, Bennett et al. [7] reported that elevated urinary NGAL levels occurring as early as 2 h after surgery predicted the severity and duration of AKI, prolonged length of stay, RRT requirement, and death. Serial measurements of urinary kidney injury molecule (KIM)-1 and IL-18 starting prior to and following cardiac surgery with cardiopulmonary bypass (CPB) were correlated with both the diagnosis and likelihood of worsening AKI [8]. Other biomarkers that have been shown to be associated with worsening severity of AKI include plasma cystatin C and urinary π-GST. A prospective study of 123 adults undergoing cardiac surgery found urinary π-GST levels at the time of increase in serum creatinine predicted worsening AKI and progression to AKIN stage 3 [9].

In settings where the measurement of selected biomarkers are made at the time of hospital or ICU admission, and hence may not be measured in relative proximity to the time of injury, biomarkers still appear to have some prognostic utility for adverse outcomes. In a study of more than 500 critically ill adult patients, urinary NGAL, cystatin C, and IL-18 measurements at the time of ICU admission predicted not only AKI, but also worsening AKI and need for RRT [10]. Of interest, in this study the association with adverse outcomes such as hospital mortality was stronger than for AKI itself. In another study of 451 critically ill patients, urinary IL-18 measured within 24 h of ICU admission was a strong predictor of death and/or receipt of acute RRT within 28 days. Similar to the previous study, urinary IL-18 predicted hospital mortality but not the development of AKI [11]. Urinary NGAL has also been shown to predict a composite outcome that included worsening severity of AKI in hospitalized patients with established AKI [12]. Furthermore, semiquantitative urinary sediment scoring has been shown to have some value in the prognostication of AKI in hospitalized patients receiving a nephrology consult [13]. Development and standardization of urinary sediment analysis are ongoing, but no consensus on standardized processing and scoring of urinary sediment currently exist. Hall et al. [14] assessed 249 patients, half of which were critically ill, and compared several novel AKI biomarkers with urine sediment examination to predict adverse outcomes. These data were consistent with additional reports suggesting urine sediment may have added diagnostic and prognostic value to novel AKI biomarkers [15].

In addition, the value of urine output as a biomarker in AKI has recently been reappraised, as there has been incomplete understanding and controversy regarding the precise utility of oliguria in the diagnosis and prognosis of AKI. To understand this further, Macedo et al. [16] utilized high-fidelity urimeters to compare various definitions of oliguria in critically ill patients. The observed incidence of AKI increased significantly from 24 to 52% when adding the urine output criteria, as defined by the AKIN classification system, to the serum creatinine-based criteria alone. When the most conservative measure of oliguria was assessed, patients without a change in serum creatinine had ICU mortality similar to oliguric patients with an increase in serum creatinine (8.8 vs. 10.4%).

Moreover, oliguria was a more sensitive marker of AKI, and tended to occur earlier than did change in serum creatinine. These data reaffirm the clinical relevance and importance of the urine output criteria to the AKIN classification system for AKI.

In terms of blood biomarkers, in two large general adult ICU populations, elevated plasma NGAL levels at ICU admission showed very good discrimination of patients who subsequently received RRT, with areas under the receiver operator curve (AuROC) of 0.82–0.88, however, showed only fair prediction of mortality (AuROCs 0.63–0.68) [17, 18]. Plasma NGAL levels also correlated with severity and duration of AKI, as well as general severity of illness scores including Acute Physiology and Chronic Health Evaluation II (APACHE II) [17]. Furthermore, in selected ICU patients with severe AKI treated with RRT, plasma NGAL levels at the time of RRT initiation were higher in non-survivors compared to survivors, and were independently associated with 28-day survival [19]. Serum cystatin C has also been reported to be an independent predictor of the severity of AKI after cardiac surgery [20], but prediction of RRT use or composite of RRT and death in ICU or nephrology consult populations have been more modest (AuROCs 0.62–0.76) [21–23].

With regard specifically to RRT, early initiation of renal support in selected patients with established AKI may have the potential to improve outcomes [24]. If novel biomarkers were able to predict the need for RRT, they could potentially be used to inform the decision to initiate RRT. A recent review, examining NGAL, cystatin C, NAG, KIM-1, and α_1-microglobulin, among others, suggested that selected novel biomarkers have the potential to discriminate earlier in those who will develop worsening AKI and need RRT [25]. However, these data require additional verification due to several limitations, including variation across studies on when RRT is initiated, which results in heterogeneity of timing of biomarker measurement and associated variability in the prognostic predictive value of these measurements [25].

At the moment, few studies have addressed the utility of biomarkers for predicting the timing and extent of kidney recovery (if any) after an episode of AKI. The recovery process depends on several factors, including the severity of the initial injury and whether there are ongoing insults. In a substudy of 76 critically ill non-anuric patients from the Acute Renal Failure Trial Network (ATN), Srisawat et al. [26] examined whether serial changes in urine biomarkers identified those RRT-requiring AKI patients who recovered (i.e. recovered renal function and were alive at 60 days). While a composite of biomarkers alone, including urinary cystatin C, hepatocyte growth factor, and NGAL achieved AuROC estimates of ~0.7–0.8 for predicting kidney recovery, improvement in discrimination (AuROC >0.9) was obtained only after integrating clinical parameters, such Charlson comorbidity index and APACHE II scores, into the clinical prediction model. The authors concluded that a panel of urine biomarkers can improve clinical risk prediction for non-recovery after AKI. This study highlights a few points. First, given the competing risk of renal recovery and death, it comes as no surprise that when 'recovery' requires both, additional clinical parameters would have to be added to the models to improve their precision. Second, serial measurements of at least two urinary biomarkers are needed to improve precision. Third, despite the heterogeneous precipitants for AKI, these urinary biomarkers appear to have broad prognostic utility. While this study had a small sample size and needs further verification, it highlights an important consideration. An ideal biomarker for a particular disease, such as AKI, should incrementally add value in addition to the baseline clinical risk assessment.

Lastly, in a study on acute heart failure (HF), discharge plasma NGAL level was a stronger predictor of 30-day outcome, including re-hospitalization or death, than brain natriuretic peptide (BNP), the guideline-supported standard for risk prediction, even after adjustment for discharge kidney function [27]. One could speculate that a high NGAL at the time of discharge following treatment for an episode of acute HF might imply a greater need for surveillance and monitoring following hospital discharge.

Cut-Off/Threshold Values for Prognostication in AKI

For the prediction of a clinical outcome, such as worsening AKI, need for RRT or death, two cut-offs would ideally provide the most useful information for clinicians [25]. The first cut-off would have a high negative predictive value (NPV) and indicate a low probability of the event (fig. 1). The second cut-off would have a high positive predictive value (PPV) and imply an increased probability of the event. Values in between these two thresholds would constitute an area of less certainty, a so-called 'grey area'. This is analogous to the approach currently taken with other biomarkers already in clinical use (i.e. BNP in heart failure), and could potentially aid clinical decision-making for patients with AKI (i.e. timing of initiation of RRT) [25]. For example, a low NPV cut-off would identify AKI patients with absent or mild tubular injury who are less likely to have worsening AKI and would be expected to have early recovery of kidney function. Alternatively, a high PPV cut-off would identify early those patients who have more severe kidney injury and are at increased risk for persistent and/or worsening AKI, in whom earlier commencement of RRT may be a consideration. For patients with values in the 'grey area', additional clinical factors can be integrated and/or serial biomarker evaluation performed to better clarify prognosis and assist clinical decision-making.

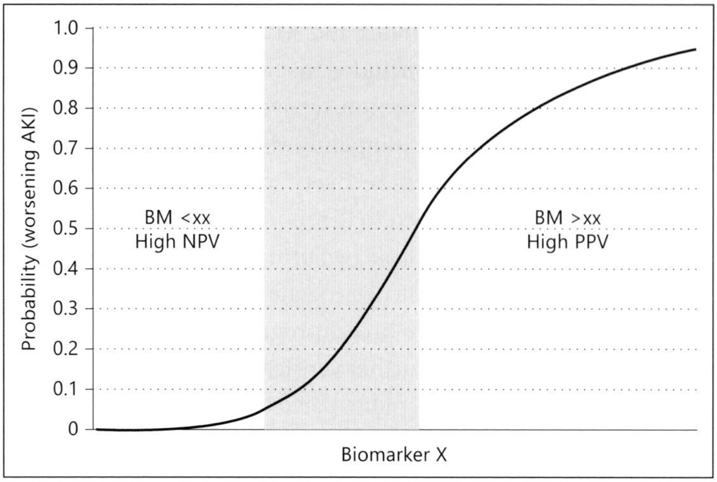

Fig. 1. Biomarker context-specific thresholds: for the prediction of a clinical outcome, e.g. worsening AKI, two cut-offs provide useful information [25]. The first cut-off would have a high NPV, and indicates a low probability of the event. The second cut-off would have a high PPV, and represents a high probability of the event. Values in between these two cut-offs represent an area of prognostic uncertainty, the so-called 'grey area'. Biomarkers may show variable operative characteristics when stratified by certain conditions, for example the high NPV and high PPV cut-off points may be shifted to the right in sepsis. Reproduced with permission from ADQI [58].

There remain important challenges to adopting novel biomarkers into clinical practice for the diagnosis and prognosis in AKI. At present, few studies have reported specific cut-off values in the above manner. Indeed, some studies have a conceptual approach to the analysis for prediction of RRT. While analysis by 'per 1-SD increase' [22] or by quartiles [28] or quintiles [29] do confirm that a higher biomarker level is more predictive of clinical events such as worsening AKI, need for RRT or death, it has not yet translated into consistent clinically meaningful cut-off values which can be pragmatically applied at the bedside. Second, a limitation of these studies is the variable timing and frequency of specimen collection. Because different biomarkers may have varying kinetics following AKI, timing of specimen collection with respect to the relevant kidney insult may significantly unduly influence their predictive potential. Furthermore, some studies examine the utility of a single measurement, while others examine prognosis related to serial biomarker measurements. Moreover, cut-off values may be specific to certain clinical contexts (discussed further below). Finally, technical differences across biomarker assays, including issues of reliability and accuracy, pose additional challenges for the practicing clinician looking for specific threshold values. The availability of commercial clinical laboratory platforms for biomarker measurement is expected to contribute to analytical harmonization, as has been done for the natriuretic peptides (NP) in HF and serum creatinine for estimation of glomerular filtration rate (eGFR) [30].

Based on the clinical context, there is sufficient data to suggest selected functional and damage biomarkers can be used to predict clinical outcomes, including worsening AKI, RRT utilization, mortality, renal recovery and a composite of these. However, there are knowledge gaps in current understanding of thresholds for the interpretation of biomarker values to predict clinical outcomes. Additional research must clarify the optimal biomarker values/cut-points for prognostication and for informing clinical management decisions across all clinically relevant outcomes, particularly with combinations of functional and damage markers.

Factors Affecting the Operative Characteristics of Biomarkers for Prognostication in AKI

Data has increasingly accumulated on several kidney injury biomarkers suggesting that their diagnostic and prognostic performance may vary depending on the clinical setting in which they are evaluated [31]. These observed differences in biomarker performance, across demographic features, co-morbidities and clinical settings have created unique challenges for interpretation and generalizability. These findings suggest that we need to clarify if and to what extent these biomarkers show variable patterns of expression (static and dynamic) across standard demographic features such as age and sex, analogous to age- and sex-specific normative data for conventional measures of kidney function (i.e. serum creatinine). The heterogeneity of patient populations, covariates used in multivariable modeling, assay differences, and need for cost-benefit analyses all preclude uniform adaptation of biomarkers into routine clinical practice at the present time.

CKD is a potentially important modifier of the performance of biomarkers for diagnosis/prognosis in AKI [10, 11, 32]. McIlroy et al. [32] found urine NGAL was a better diagnostic predictor of AKI, defined as an >50% or >0.3 mg/dl in serum creatinine, following cardiac surgery in those with a baseline eGFR ≥60 ml/min/1.73 m^2. Siew et al. found modest improvement in the performance of urine IL-18 for predicting AKI within 24 h of ICU admission when stratified by eGFR ≥60 ml/min/1.73 m^2 [11]. Endre et al. [10] also recently reported variable performance of several biomarkers for AKI diagnosis and prediction of outcomes when stratified by both eGFR and by time after kidney insult in a prospective cohort of 529 critically ill patients. At the present time however, there are no such established differential cut-off values for renal biomarkers among patients with and without CKD.

Other than CKD, there remains uncertainty whether the variable expression of these biomarkers across clinical settings represents differences attributable to isolated 'kidney injury' or response to increased 'systemic' illness. Experimental models of ischemia-reperfusion injury have clearly improved our understanding of the specific kidney expression of biomarkers such as NGAL [33], however there is uncertainty of how its performance will extend to more general clinical settings. This may account for the inconsistent performance between biomarkers evaluated in relatively isolated settings of AKI, such as cardiac surgery-associated AKI (post-CPB), contrast nephropathy, or delayed graft function after kidney transplantation, versus more complex multisystem illness settings in which numerous additional confounders may be present (i.e. sepsis, malignancy) [31]. This would imply there is need for added understanding of not only 'context-specific' performance, but also their time-specific (temporal patterns), and illness severity-specific (non-kidney organ injury) performance. While these biomarkers clearly show correlation with AKI, in particular in models of ischemia-reperfusion and nephrotoxicity, selected biomarkers including NGAL [17, 34, 35], IL-18 [36], L-FABP [37], can show increased expression with systemic illness and non-kidney organ injury. Moreover, we do not understand how an increasing severity of kidney injury stimulus can induce secondary organ dysfunction (i.e. kidney-organ crosstalk), and may further bias its operative performance. For example, increased biomarker expression in septic AKI compared with non-septic AKI patients has been shown in some studies [11, 18, 19, 38, 39], however not in others [17]. In other words, the high NPV and high PPV cut-off points may be shifted to the right in sepsis (fig. 1). It is unknown how the phase of sepsis (i.e. acute, subacute, recovering) might impact this phenomenon. Moreover, selected data further suggest the diagnostic and prognostic performance may improve if septic patients are excluded [38, 40].

In summary, it is important that clinicians who intend to use biomarkers in clinical practice be aware that selected biomarkers show variable operative characteristics when stratified by CKD and sepsis. Additional research on selected functional and damage biomarkers is needed for independent validation of across demographic features, selected comorbidities (in particular CKD), and across clinical settings (in particular sepsis) and illness severity.

Serial Biomarker Measurements for Prognostication in AKI

There remains uncertainty on whether serial biomarker evaluation in patients at-risk for or early after developing AKI can provide additional information for clarifying prognosis.

There have been numerous studies that have described the temporal patterns of biomarker detection in AKI, including after cardiac surgery [7, 8, 29, 41–45] and in critically ill or septic patients [17–19, 38, 46]. However, few studies have provided any detailed description of the potential added prognostic value of evaluating the rate of or time-averaged change, the absolute or relative magnitude of change (i.e. delta), or the trajectory of change (i.e. increasing, decreasing, stable) of selected damage biomarkers after the occurrence of a kidney insult, compared with serial conventional measure of kidney function such as serum creatinine. Changes in urine output and patterns of change in serum creatinine have been shown to have prognostic information representing alterations in functional biomarkers [16, 47], however these changes have not been correlated with concurrent alterations in damage markers. Moreover, no study has evaluated the potential additive value of serial biomarker evaluation in response to therapeutic interventions such as acute resuscitation, hemodynamic optimization, antimicrobial therapy in sepsis, withdrawal of nephrotoxins, and other novel interventions in the setting of a randomized trial. Such data would be important in determining whether such biomarkers can be used for 'bio-monitoring' (see below). In a prospective study of 632 consecutive critically ill patients, de Geus et al. [18] evaluated the diagnostic and prognostic performance of serial plasma and urine NGAL measures early after ICU admission. In this secondary analysis, temporal changes in plasma or urine NGAL during the first 24 h did not significantly improve the prediction of development of RIFLE-F, rather the most proximate measure of NGAL temporally associated with achieving RIFLE-F was more predictive.

Serial biomarker evaluation has also been studied for clinical prediction of recovery of kidney function following an episode of AKI. In the above-mentioned substudy of the ATN trial, the addition of the largest relative changes across a panel of urine biomarkers at 14 days, when combined with the most robust clinical risk prediction model, significantly improved overall model performance for predicting kidney recovery [26].

In theory, the use of serial measures of selective biomarkers may have added clinical importance in settings where the time of kidney insult is uncertain and/or where there are multiple temporally separated contributing risks for AKI [10, 48]. The timing of biomarker measurement relative to kidney insult appears to significantly impact its performance for the prediction and/or diagnosis of AKI [10]. Likewise, we need improved understanding of the variability in pathophysiologic expression of kidney injury-specific biomarkers relative to the kidney insult for detecting discrete 'phases' of injury that correlate not only with severity and duration of injury, but also with loss of function and/or clinically important endpoints. It has been postulated that differences in performance for detection of early, sustained or resolving kidney damage may relate to whether the biomarker is preformed and constitutive (i.e. tubular brush border enzymes)

Table 2. New definition of the ideal biomarker

Either highly sensitive (diagnosis) *or* highly specific (treatment effect)
Reflects abnormal physiology/biochemistry
Prognosis is most meaningful if level is clinically actionable
Should be used as a basis for specific 'biomarker guided therapy'
'Bio-monitoring' during treatment is an effective surrogate of improvement

or inducible and upregulated (i.e. NGAL, KIM-1) in the presence of tubular injury. Furthermore, serial biomarker measurements could potentially identify the occurrence of new insults, however studies which address this issue directly are currently lacking.

Clearly, serial biomarker measurements could potentially add to healthcare costs. In addition to the knowledge gaps already described above, there are no data specifically evaluating the cost-to-benefit ratio of serial measurement in the prognostication of AKI.

Overall, while we believe temporal assessment of biomarkers is likely to have value for clarifying prognostic uncertainty, we are unable to recommend a specific serial testing schedule based on the available literature. Further studies are needed to clarify the role of serial measures of selected biomarkers for the diagnosis and projected severity of AKI, clarifying prognostic uncertainty, and for the assessment of kidney-specific response to interventions.

Biomarkers and Clinical Management Decisions

Until recently, most clinicians were satisfied with the definition of an ideal biomarker as a sensitive and specific tool for diagnosis of a disease state as well as prognostic for near-term clinical outcomes [49]. More recently, a new updated definition of the ideal biomarker has been proposed (table 2) [50]. In this paradigm, a 'prognostic' biomarker is only important as a risk-stratifying tool if it is 'clinically actionable' (i.e. it can be used to start a clinically relevant therapy and/or can be used to 'bio-monitor' the treatment success or failure).

The early identification of AKI by biomarkers may be particularly 'actionable' in the emergency room (ER) setting, where the potential adverse consequences of sending home a patient with evolving AKI could be severe. A reduction in urine flow has been demonstrated to be an early physiologic marker of renal functional change associated with adverse outcomes, particularly when measured accurately [16]. In both adult and ER pediatric settings, NGAL has been shown to identify patients with AKI and who proceed to require higher intensity monitoring and support, such as ICU admission, nephrology consultation and initiation of RRT [51, 52]. In this context, it can be useful for clinical triage of patients to either hospital admission, or at the very least, further evaluation in the ER, perhaps by a nephrologist. In an analogous manner, this could potentially be applied to the ward setting, in that patients with positive biomarkers can be moved to a higher acuity unit with enhanced monitoring capacity. Although this approach has not been tested in a randomized controlled manner, it has potential benefits with (debatably) relatively little foreseeable clinical 'downside'. However, it is acknowledged that monitoring a large number of patients, some of whom will be 'false positives', would impact on healthcare costs. This aspect deserves further study.

At least two studies have shown that delayed nephrology consultation can be associated with increased ICU mortality [53, 54]. Lower serum creatinine level and higher urine output were associated with delayed consultation, suggesting that these traditional biomarkers may have led to underestimation of the severity of kidney dysfunction in these patients. In a separate prospective controlled non-randomized intervention study, a compulsory one-time nephrology consult was performed within 18 h of fulfilling AKIN criteria (increase in sCr ≥0.3 mg/dl) for AKI [55]. In this single-center study, this prompt nephrology consultation was associated with a lower peak serum creatinine. The primary outcome of a 2.5-fold increase in serum creatinine occurred in 12.9% of patients in the control group compared with 3.3% of patients who received early nephrology consultation. In another article in this volume it has been proposed that novel biomarkers be integrated with RIFLE/AKIN for the diagnosis of AKI, specifically that abnormal levels of biomarkers may be used to diagnose AKI even in the absence of changes in serum creatinine or oliguria. Patients with elevated biomarkers without diagnostic changes in serum creatinine are at higher risk for adverse outcomes, including need for RRT and mortality [56]. Although not yet supported by RCT evidence, we suggest that consultation with clinicians with AKI expertise should be considered in response to elevation in functional and/or damage biomarkers.

In the absence of specific therapeutic interventions, it may seem curious that the early introduction of nephrology expertise seems to improve outcomes. We believe that this is likely due earlier identification of the important of AKI, consideration for its etiology, and withdrawal or appropriate dose adjustment of potential nephrotoxins, many of which have only been recently identified. In the above-mentioned study [55], the observed effect on peak serum creatinine was achieved with fairly simple diagnostic and therapeutic recommendations (table 3). In another article in this volume it has been recommended that the etiology of AKI should be determined as soon as the diagnosis is made volume. Context-specific biomarkers (i.e. natriuretic peptides, procalcitonin, renal imaging) may assist in determining the etiology and severity of AKI, and could be per-

Table 3. Potential clinical actions in response to altered biomarker levels

Triage decisions
Examples: admitted from ER, hold for further observation, move from ward to higher acuity unit
Nephrology consultation
Differential diagnosis
Search for reversible hemodynamic causes
Search for potential direct nephrotoxins
Context-specific biomarkers[1], imaging and other studies to establish AKI etiology
Renoprotective strategies
Higher acuity monitoring (fluid balance, urine output, creatinine, blood pressure, cardiac function)
Hold ACE-I/ARB
Optimize volume status and perfusion pressure
Adjust diuretic doses
Initial use of vasodilators, including nitrates, hydralazine and nesiritide (in acute heart failure)
Pharmacovigilance (drug monitoring/dosing, avoiding nephrotoxins, attention to drug interaction)
Radiocontrast precautions/consider alternatives to radiocontrast procedures
Supportive management
Avoid hyperglycemia
Anticipate AKI complications (e.g. hyperkalemia, volume overload, etc.)
Consider adjusting nutrition (e.g. attention to K, PO_4 content, volume, etc.)
Consider timely initiation of renal support
Avoid subclavian catheters if possible
Context-specific intervention(s)
Consider early or closer follow-up

[1] Examples of context-specific biomarkers: BNP and NT-proBNP for heart failure; bilirubin and hepatic enzymes for hepatic failure; procalcitonin, endotoxin activity assay and cultures for sepsis.

formed in response to abnormal biomarker levels. Furthermore, renal protective strategies, as listed in table 3, can be easily implemented. Although the beneficial effect of these interventions was shown in a study in which the trigger was a change in serum creatinine, we suggest that these actions can be undertaken at the time of AKI diagnosis, be it by serum creatinine, urine output or biomarker.

Based on the clinical context, there are data to suggest selected biomarkers can be important tools to aid in patient triage and management decisions (e.g. renal protective strategies, nephrology consultation). Biomarker assays have been introduced in some countries but their clinical uptake has not been high enough to develop a knowledge base of thresholds, however based on published observational studies of these biomarkers, potential thresholds can be suggested but will need to be validated. Additional research must clarify the optimal bio-

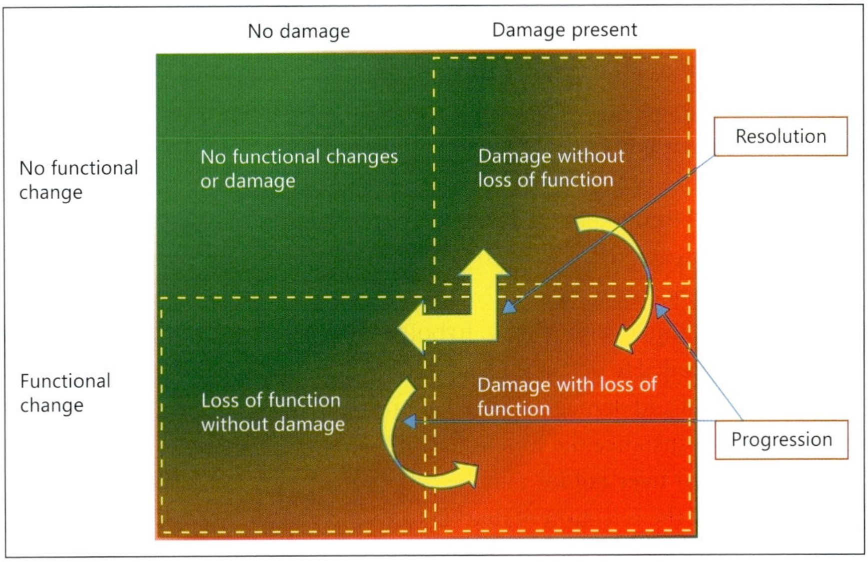

Fig. 2. Utilization of functional and damage markers for prognosis in AKI. The combined use of functional and damage biomarkers permits categorization of AKI patients into one of the four quadrants shown at initial evaluation. Based on this distinction, serial measurements can be used to determine the clinical relevance of each category for prognosis for renal (e.g. recovery of kidney function, need for RRT) and non-renal (e.g. mortality, non-renal organ dysfunction, complications). Sequential assessments could provide information on transitions among the quadrants and determine which of the factors is prevalent for ongoing injury or resolution and offer opportunities for targeted intervention. Future studies would be needed to validate the prognostic information of these groupings across the spectrum of AKI and across various disease states. Reproduced with permission from ADQI [58].

marker values for categorizing high PPV/NPV for specific clinical management decisions. Furthermore, research on cost-benefit assessment of such biomarker-guided clinical action plans is needed.

Conclusion

Over the past decade, a multitude of AKI damage biomarkers have been developed and are now being brought into clinical practice. We anticipate that more AKI functional and damage biomarkers will be emerging alongside the ones that are currently available. As discussed earlier and shown in figure 2, the combination of functional and damage biomarkers provides a novel set of tools for the clinician to manage patients with AKI. At initial evaluation, patients can be categorized into one of the four diagnostic categories that each would have potential clinical significance with respect to outcomes. For instance, patients in the functional or damage-alone categories may progress to a stage of combined events or resolve to a state of normalcy. Consequently, recognition that the patients are in one of these phases of injury and/or loss of function could permit targeted intervention to prevent further progression. Serial measurements with the biomarkers would allow determination of transitions between these diagnostic categories and provide further information on the prognosis. They could also be indicated for 'dynamic' changes in clinical status or following subsequent 'injury events' (e.g. hypotension). It is evident that further research will be needed to determine the validity of these concepts and also define which functional and damage biomarkers are best suited based on the underlying context. For instance the functional and damage markers for patients with AKI in setting of liver failure may be different than those in setting of CKD both in terms of the selection of markers and their thresholds. We believe that when used together, functional and damage markers can be used to assist in decisions related to triage of patients with AKI and identifying patients with who are at risk for progression. Set cut-offs for various biomarkers and their bedside utility are forthcoming and will be in part determined by regulatory intended use guidelines, platform standardization, and inter-laboratory calibration. There remain many unresolved areas of AKI biomarker use in selected syndromes of AKI (e.g. cardiorenal syndrome, hepatorenal syndrome). Prospective studies utilizing combinations of AKI functional and damage biomarkers in a fashion that directly impact patients' care in these areas is required. We expect that as clinicians gain experience with AKI biomarkers, clinical care plans that incorporate them into routine care will shortly follow.

References

1 Fischer MJ, Brimhall BB, Lezotte DC, Glazner JE, Parikh CR: Uncomplicated acute renal failure and hospital resource utilization: a retrospective multicenter analysis. Am J Kidney Dis 2005;46:1049–1057.

2 Bagshaw SM, Laupland KB, Doig CJ, Mortis G, Fick GH, Mucenski M, Godinez-Luna T, Svenson LW, Rosenal T: Prognosis for long-term survival and renal recovery in critically ill patients with severe acute renal failure: a population-based study. Crit Care 2005; 9:R700–R709.

3 Ponte B, Felipe C, Muriel A, Tenorio MT, Liano F: Long-term functional evolution after an acute kidney injury: a 10-year study. Nephrol Dial Transplant 2008;23:3859–3866.

4 Cruz DN, Ronco C: Acute kidney injury in the intensive care unit: current trends in incidence and outcome. Crit Care 2007;11: 149.

5 Chawla LS, Amdur RL, Amodeo S, Kimmel PL, Palant CE: The severity of acute kidney injury predicts progression to chronic kidney disease. Kidney Int 2011;79:1361–1369.

6 Parikh CR, Devarajan P, Zappitelli M, Sint K, Thiessen-Philbrook H, Li S, Kim RW, Koyner JL, Coca SG, Edelstein CL, Shlipak MG, Garg AX, Krawczeski CD: Postoperative biomarkers predict acute kidney injury and poor outcomes after pediatric cardiac surgery. J Am Soc Nephrol 2011;22:1737–1747.

7 Bennett M, Dent CL, Ma Q, Dastrala S, Grenier F, Workman R, Syed H, Ali S, Barasch J, Devarajan P: Urine NGAL predicts severity of acute kidney injury after cardiac surgery: a prospective study. Clin J Am Soc Nephrol 2008;3:665–673.

8 Liangos O, Tighiouart H, Perianayagam MC, Kolyada A, Han WK, Wald R, Bonventre JV, Jaber BL: Comparative analysis of urinary biomarkers for early detection of acute kidney injury following cardiopulmonary bypass. Biomarkers 2009;14:423–431.

9 Koyner JL, Vaidya VS, Bennett MR, Ma Q, Worcester E, Akhter SA, Raman J, Jeevanandam V, O'Connor MF, Devarajan P, Bonventre JV, Murray PT: Urinary biomarkers in the clinical prognosis and early detection of acute kidney injury. Clin J Am Soc Nephrol 2010;5:2154–2165.

10 Endre ZH, Pickering JW, Walker RJ, Devarajan P, Edelstein CL, Bonventre JV, Frampton CM, Bennett MR, Ma Q, Sabbisetti VS, Vaidya VS, Walcher AM, Shaw GM, Henderson SJ, Nejat M, Schollum JB, George PM: Improved performance of urinary biomarkers of acute kidney injury in the critically ill by stratification for injury duration and baseline renal function. Kidney Int 2011;79:1119–1130.

11 Siew ED, Ikizler TA, Gebretsadik T, Shintani A, Wickersham N, Bossert F, Peterson JF, Parikh CR, May AK, Ware LB: Elevated urinary IL-18 levels at the time of ICU admission predict adverse clinical outcomes. Clin J Am Soc Nephrol 2010;5:1497–1505.

12 Singer E, Elger A, Elitok S, Kettritz R, Nickolas TL, Barasch J, Luft FC, Schmidt-Ott KM: Urinary neutrophil gelatinase-associated lipocalin distinguishes pre-renal from intrinsic renal failure and predicts outcomes. Kidney Int 2011;80:405–414.

13 Perazella MA, Coca SG, Hall IE, Iyanam U, Koraishy M, Parikh CR: Urine microscopy is associated with severity and worsening of acute kidney injury in hospitalized patients. Clin J Am Soc Nephrol 2010;5:402–408.

14 Hall IE, Coca SG, Perazella MA, Eko UU, Luciano RL, Peter PR, Han WK, Parikh CR: Risk of poor outcomes with novel and traditional biomarkers at clinical AKI diagnosis. Clin J Am Soc Nephrol 2011;6:2740–2749.

15 Perazella MA, Coca SG: Traditional urinary biomarkers in the assessment of hospital-acquired AKI. Clin J Am Soc Nephrol 2012;7:167–174.

16 Macedo E, Malhotra R, Bouchard J, Wynn SK, Mehta RL: Oliguria is an early predictor of higher mortality in critically ill patients. Kidney Int 2011;80:760–767.

17 Cruz DN, de Cal M, Garzotto F, Perazella MA, Lentini P, Corradi V, Piccinni P, Ronco C: Plasma neutrophil gelatinase-associated lipocalin is an early biomarker for acute kidney injury in an adult ICU population. Intensive Care Med 2010;36:444–451.

18 de Geus HR, Bakker J, Lesaffre EM, le Noble JL: Neutrophil gelatinase-associated lipocalin at ICU admission predicts for acute kidney injury in adult patients. Am J Respir Crit Care Med 2011;183:907–914.

19 Kumpers P, Hafer C, Lukasz A, Lichtinghagen R, Brand K, Fliser D, Faulhaber-Walter R, Kielstein JT: Serum neutrophil gelatinase-associated lipocalin at inception of renal replacement therapy predicts survival in critically ill patients with acute kidney injury. Crit Care 2010;14:R9.

20 Haase M, Bellomo R, Devarajan P, Ma Q, Bennett MR, Mockel M, Matalanis G, Dragun D, Haase-Fielitz A: Novel biomarkers early predict the severity of acute kidney injury after cardiac surgery in adults. Ann Thorac Surg 2009;88:124–130.

21 Herget-Rosenthal S, Marggraf G, Husing J, Goring F, Pietruck F, Janssen O, Philipp T, Kribben A: Early detection of acute renal failure by serum cystatin C. Kidney Int 2004;66:1115–1122.

22 Perianayagam MC, Seabra VF, Tighiouart H, Liangos O, Jaber BL: Serum cystatin C for prediction of dialysis requirement or death in acute kidney injury: a comparative study. Am J Kidney Dis 2009;54:1025–1033.

23 Nejat M, Pickering JW, Walker RJ, Endre ZH: Rapid detection of acute kidney injury by plasma cystatin C in the intensive care unit. Nephrol Dial Transplant 2010;25:3283–3289.

24 Seabra VF, Balk EM, Liangos O, Sosa MA, Cendoroglo M, Jaber BL: Timing of renal replacement therapy initiation in acute renal failure: a meta-analysis. Am J Kidney Dis 2008;52:272–284.

25 Cruz DN, de Geus HR, Bagshaw SM: Biomarker strategies to predict need for renal replacement therapy in acute kidney injury. Semin Dial 2011;24:124–131.

26 Srisawat N, Wen X, Lee M, Kong L, Elder M, Carter M, Unruh M, Finkel K, Vijayan A, Ramkumar M, Paganini E, Singbartl K, Palevsky PM, Kellum JA: Urinary biomarkers and renal recovery in critically ill patients with renal support. Clin J Am Soc Nephrol 2011;6:1815–1823.

27 Maisel AS, Mueller C, Fitzgerald R, Brikhan R, Hiestand BC, Iqbal N, Clopton P, van Veldhuisen DJ: Prognostic utility of plasma neutrophil gelatinase-associated lipocalin in patients with acute heart failure: the NGAL EvaLuation Along with B-type NaTriuretic Peptide in acutely decompensated heart failure (GALLANT) trial. Eur J Heart Fail 2011; 13:846–851.

28 Liangos O, Perianayagam MC, Vaidya VS, Han WK, Wald R, Tighiouart H, MacKinnon RW, Li L, Balakrishnan VS, Pereira BJ, Bonventre JV, Jaber BL: Urinary N-acetyl-β-D-glucosaminidase activity and kidney injury molecule-1 level are associated with adverse outcomes in acute renal failure. J Am Soc Nephrol 2007;18:904–912.

29 Parikh CR, Devarajan P, Zappitelli M, Sint K, Thiessen-Philbrook H, Li S, Kim RW, Koyner JL, Coca SG, Edelstein CL, Shlipak MG, Garg AX, Krawczeski CD: Postoperative biomarkers predict acute kidney injury and poor outcomes after adult cardiac surgery. J Am Soc Nephrol 2011;22:1748–1757.

30 Levey AS, Coresh J, Greene T, Stevens LA, Zhang YL, Hendriksen S, Kusek JW, Van Lente F: Using standardized serum creatinine values in the modification of diet in renal disease study equation for estimating glomerular filtration rate. Ann Intern Med 2006;145: 247–254.

31 Siew ED, Ware LB, Ikizler TA: Biological markers of acute kidney injury. J Am Soc Nephrol 2011;22:810–820.

32 McIlroy DR, Wagener G, Lee HT: Neutrophil gelatinase-associated lipocalin and acute kidney injury after cardiac surgery: the effect of baseline renal function on diagnostic performance. Clin J Am Soc Nephrol 2010;5:211–219.

33 Paragas N, Qiu A, Zhang Q, Samstein B, Deng SX, Schmidt-Ott KM, Viltard M, Yu W, Forster CS, Gong G, Liu Y, Kulkarni R, Mori K, Kalandadze A, Ratner AJ, Devarajan P, Landry DW, D'Agati V, Lin CS, Barasch J: The Ngal reporter mouse detects the response of the kidney to injury in real time. Nat Med 2011;17:216–222.

34 Eagan TM, Damas JK, Ueland T, Voll-Aanerud M, Mollnes TE, Hardie JA, Bakke PS, Aukrust P: Neutrophil gelatinase-associated lipocalin: a biomarker in COPD. Chest 2010; 138:888–895.

35 Marti J, Fuster J, Hotter G, Sola AM, Deulofeu R, Modolo MM, Loera MA, Ferrer J, Fondevila C, Garcia-Valdecasas JC: Serum neutrophil gelatinase-associated lipocalin in patients with colorectal liver metastases: preliminary results of an exploratory prospective study. Int J Biol Markers 2010;25:21–26.

36 Parikh CR, Abraham E, Ancukiewicz M, Edelstein CL: Urine IL-18 is an early diagnostic marker for acute kidney injury and predicts mortality in the intensive care unit. J Am Soc Nephrol 2005;16:3046–3052.

37 Lachmann RA, Werchan S, Schachtrup C, Haitsma JJ, Spener F, Lachmann B: Liver-type fatty acid binding protein in serum and bronchoalveolar lavage in a model of acute respiratory failure because of surfactant depletion – a possible marker for lung damage? Clin Physiol Funct Imaging 2006;26:371–375.

38 Bagshaw SM, Bennett M, Haase M, Haase-Fielitz A, Egi M, Morimatsu H, D'Amico G, Goldsmith D, Devarajan P, Bellomo R: Plasma and urine neutrophil gelatinase-associated lipocalin in septic versus non-septic acute kidney injury in critical illness. Intensive Care Med 2010;36:452–461.

39 Martensson J, Bell M, Oldner A, Xu S, Venge P, Martling CR: Neutrophil gelatinase-associated lipocalin in adult septic patients with and without acute kidney injury. Intensive Care Med 2010;36:1333–1340.

40 Washburn KK, Zappitelli M, Arikan AA, Loftis L, Yalavarthy R, Parikh CR, Edelstein CL, Goldstein SL: Urinary interleukin-18 is an acute kidney injury biomarker in critically ill children. Nephrol Dial Transplant 2008;23: 566–572.

41 Parikh CR, Coca SG, Thiessen-Philbrook H, Shlipak MG, Koyner JL, Wang Z, Edelstein CL, Devarajan P, Patel UD, Zappitelli M, Krawczeski CD, Passik CS, Swaminathan M, Garg AX: Postoperative biomarkers predict acute kidney injury and poor outcomes after adult cardiac surgery. J Am Soc Nephrol 2011;22:1748–1757.

42 Han WK, Waikar SS, Johnson A, Betensky RA, Dent CL, Devarajan P, Bonventre JV: Urinary biomarkers in the early diagnosis of acute kidney injury. Kidney Int 2008;73:863–869.

43 Haase M, Bellomo R, Story D, Davenport P, Haase-Fielitz A: Urinary interleukin-18 does not predict acute kidney injury after adult cardiac surgery: a prospective observational cohort study. Crit Care 2008;12:R96.

44 Haase-Fielitz A, Bellomo R, Devarajan P, Story D, Matalanis G, Dragun D, Haase M: Novel and conventional serum biomarkers predicting acute kidney injury in adult cardiac surgery – a prospective cohort study. Crit Care Med 2009;37:553–560.

45 Haase M, Bellomo R, Devarajan P, Ma Q, Bennett MR, Mockel M, Matalanis G, Dragun D, Haase-Fielitz A: Novel biomarkers early predict the severity of acute kidney injury after cardiac surgery in adults. Ann Thorac Surg 2009;88:124–130.

46 Zappitelli M, Washburn KK, Arikan AA, Loftis L, Ma Q, Devarajan P, Parikh CR, Goldstein SL: Urine neutrophil gelatinase-associated lipocalin is an early marker of acute kidney injury in critically ill children: a prospective cohort study. Crit Care 2007; 11:R84.

47 Prowle JR, Liu YL, Licari E, Bagshaw SM, Egi M, Haase M, Haase-Fielitz A, Kellum JA, Cruz D, Ronco C, Tsutsui K, Uchino S, Bellomo R: Oliguria as predictive biomarker of acute kidney injury in critically ill patients. Crit Care 2011;15:R172.

48 Haase M, Devarajan P, Haase-Fielitz A, Bellomo R, Cruz DN, Wagener G, Krawczeski CD, Koyner JL, Murray P, Zappitelli M, Goldstein SL, Makris K, Ronco C, Martensson J, Martling CR, Venge P, Siew E, Ware LB, Ikizler TA, Mertens PR: The outcome of neutrophil gelatinase-associated lipocalin-positive subclinical acute kidney injury: a multicenter pooled analysis of prospective studies. J Am Coll Cardiol 2011; 57:1752–1761.

49 Morrow DA, de Lemos JA: Benchmarks for the assessment of novel cardiovascular biomarkers. Circulation 2007;115:949–952.

50 Maisel A: Biomonitoring and biomarker-guided therapy: the next step in heart failure and biomarker research. J Am Coll Cardiol 2011;58:1890–1892.

51 Nickolas TL, O'Rourke MJ, Yang J, Sise ME, Canetta PA, Barasch N, Buchen C, Khan F, Mori K, Giglio J, Devarajan P, Barasch J: Sensitivity and specificity of a single emergency department measurement of urinary neutrophil gelatinase-associated lipocalin for diagnosing acute kidney injury. Ann Intern Med 2008;148:810–819.

52 Shapiro NI, Trzeciak S, Hollander JE, Birkhahn R, Otero R, Osborn TM, Moretti E, Nguyen HB, Gunnerson K, Milzman D, Gaieski DF Goyal M, Cairns CB, Kupfer K, Lee SW, Rivers EP: The diagnostic accuracy of plasma neutrophil gelatinase-associated lipocalin in the prediction of acute kidney injury in emergency department patients with suspected sepsis. Ann Emerg Med 2010; 56:52–59.e1.

53 Ponce D, Zorzenon Cde P, dos Santos NY, Balbi AL: Early nephrology consultation can have an impact on outcome of acute kidney injury patients. Nephrol Dial Transplant 2011;26:3202–3206.

54 Mehta RL, McDonald B, Gabbai F, Pahl M, Farkas A, Pascual MT, Zhuang S, Kaplan RM, Chertow GM: Nephrology consultation in acute renal failure: does timing matter? Am J Med 2002;113:456–461.

55 Balasubramanian G, Al-Aly Z, Moiz A, Rauchman M, Zhang Z, Gopalakrishnan R, Balasubramanian S, El-Achkar TM: Early nephrologist involvement in hospital-acquired acute kidney injury: a pilot study. Am J Kidney Dis 2011;57:228–234.

56 Haase M, Devarajan P, Haase-Fielitz A, Bellomo R, Cruz DN, Wagener G, Krawczeski CD, Koyner JL, Murray P, Zappitelli M, Goldstein SL, Makris K, Ronco C, Martensson J, Martling CR, Venge P, Siew E, Ware LB, Ikizler TA, Mertens PR: The outcome of neutrophil gelatinase-associated lipocalin-positive subclinical acute kidney injury a multicenter pooled analysis of prospective studies. J Am Coll Cardiol 2011;57:1752–1761.

57 Doi K, Negishi K, Ishizu T, Katagiri D, Fujita T, Matsubara T, Yahagi N, Sugaya T, Noiri E: Evaluation of new acute kidney injury biomarkers in a mixed intensive care unit. Crit Care Med 2011;39:2464–2469.

58 Acute Dialysis Quality Initiative (ADQI). www.ADQI.org (accessed January 10, 2013).

Lakhmir S. Chawla, MD
Department of Anesthesiology and Critical Care Medicine
900 23rd Street, NW, RILF, G-105
Washington, DC 20037 (USA)
E-Mail minkchawla@gmail.com

McCullough PA, Kellum JA, Mehta RL, Murray PT, Ronco C (eds): ADQI Consensus on AKI Biomarkers and Cardiorenal Syndromes. Contrib Nephrol. Basel, Karger, 2013, vol 182, pp 65–81
DOI: 10.1159/000349967

Physiological Biomarkers of Acute Kidney Injury: A Conceptual Approach to Improving Outcomes

Mark D. Okusa[a] · Bertrand L. Jaber[b] · Peter Doran[c] · Jacques Duranteau[d] · Li Yang[e] · Patrick T. Murray[f] · Ravindra L. Mehta[g] · Can Ince[h] · for the ADQI 10 Workgroup

[a]Division of Nephrology and Center for Immunity, Inflammation and Regenerative Medicine, University of Virginia, Charlottesville, Va., USA; [b]Kidney and Dialysis Research Laboratory, Division of Nephrology, Department of Medicine, St. Elizabeth's Medical Center, Tufts University School of Medicine, Boston, Mass., USA; [c]UCD Clinical Research Centre, UCD School of Medicine and Medical Science, University College Dublin, Dublin, Ireland; [d]Hôpitaux de Paris, Hôpital Bicêtre, Département d'Anesthésie-Réanimation, Université Paris Sud XI, Le Kremlin-Bicêtre, France; [e]Renal Division, Peking University First Hospital, Beijing, China; [f]Department of Medicine, University College Dublin, Dublin, Ireland; [g]Department of Medicine, University of California San Diego, San Diego, Calif., USA; [h]Department of Intensive Care Adults, Erasmus MC University Medical Centre, Rotterdam, The Netherlands

Abstract

The past 5–10 years have brought significant advances in the identification and validation of novel biochemical biomarkers in the prevention and treatment of acute kidney injury (AKI). These biochemical biomarkers remain research tools but we anticipate that soon they will be employed in clinical practice. A Consensus Conference held by the Acute Dialysis Quality Initiative (ADQI) recently reviewed the evidence, and identified gaps and a research agenda. Furthermore, at this meeting was the birth of an initiative to comprehensively identify new opportunities to characterize the physiological changes during the course of AKI based upon a conceptual framework for the detection and monitoring of renal ischemia-reperfusion injury. This framework includes a transition from monitoring physiological biomarkers of adequate renal perfusion, to pathophysiologic biomarkers of renal hypoperfusion, and finally biomarkers of kidney cell structural injury/damage. Techniques to measure physiological changes in AKI include several physiological variables that might be used in an interactive way to supplement clinical information and biochemical damage biomarkers in the diagnosis and management of AKI. This review summarizes the spectrum of physiological parameters and potential new physiolog-

ADQI 10 Workgroup members are listed in Appendix 1.

ical methods that enable identification of high-risk patients for AKI, facilitate early diagnosis, and differential diagnosis to guide therapeutic management and prognostication. Finally, we propose a research agenda for the next 5 years to facilitate the development and validation of physiological biomarkers in AKI.

The discovery and validation of novel biomarkers of kidney damage over the past decade represent major advances in the field of acute kidney injury (AKI). The investment by major sponsors and investigators demonstrates a consensus that early biomarkers of AKI are needed. These damage biomarkers proved to be sensitive and specific in a number of small clinical studies, and their utility in diagnosing and prognosticating AKI awaits confirmation in larger studies. Damage biomarkers are biochemical markers of early cellular injury that can be measured upon release from cells into the plasma or urine. However, long before ischemic cellular injury, alterations in the renal microcirculation and tissue oxygenation occur, rendering epithelial cells susceptible to injury. These pathophysiological processes represent microcirculatory dysfunction ultimately leading to AKI. In addition, renal autoregulatory mechanisms are invoked in response to renal insults and play an important role in maintaining homeostasis. The underlying state of health of the kidney and the effective 'renal reserve' are important determinants of this autoregulatory response to modify renal function in response to injury. Currently, our ability to detect these changes in kidney function or structure has been limited to measurement of urine output and changes in serum creatinine reflecting alterations in the glomerular filtration rate (GFR). These measurements represent the physiological biomarkers to track the development and course of AKI and are the current parameters utilized for diagnosis and staging of AKI. However, several additional tools are now available to ascertain physiological response to kidney injury. Coupled with the emerging damage markers, these physiological biomarkers might help identify patients at risk, and provide mechanistic information and a window for therapeutic intervention to prevent AKI. As an example, we have developed a modified description of the clinical phases of ischemic AKI [1] that incorporates physiological parameters that complement the use of biochemical biomarkers and clinical characteristics for the study of AKI (fig. 1).

Methods

The 10th Acute Dialysis Quality Initiative (ADQI) Consensus Conference on 'Acute Kidney Injury Biomarkers' held in Dublin, Ireland, in August 2011 (www.adqi.net), was attended by an international group of experts, and focused on an objective scientific review of the current literature, developing a consensus of opinion, with evidence where

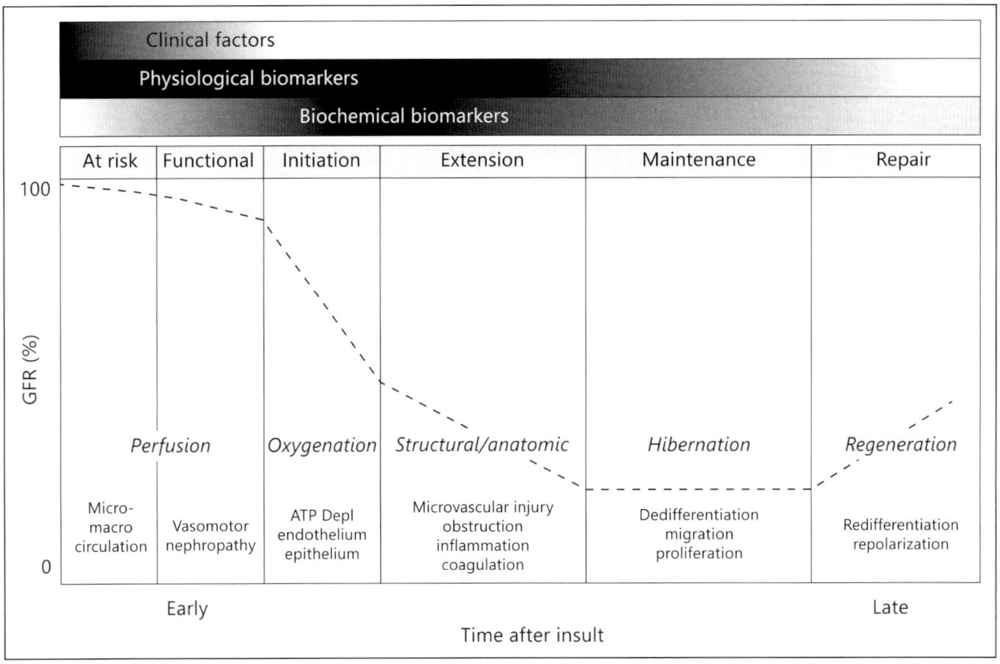

Fig. 1. Physiological phases of AKI following ischemia-reperfusion injury. The conceptual framework of physiological biomarkers is superimposed upon the previously established concept of clinical phases of AKI. This figure illustrates progression from risk to prerenal AKI and represents experimental data from ischemia reperfusion but not necessarily other forms of AKI such as drug-induced direct nephrotoxicity. Thus physiological biomarkers are not only needed in the early phase of AKI but throughout the continuum of AKI. The ability to measure these physiological variables may lead to identification of patients at risk for AKI, early diagnosis of AKI and inform therapeutic decisions. These physiological processes represent an integrative environment for the interaction of inflammatory mediators, imbalance in the homeostasis of oxygen, nitric oxide and oxygen radicals causing microcirculatory dysfunction and impaired tissue oxygenation leading to AKI. Reproduced with permission from ADQI [58].

possible, on best practice, and articulating a research agenda to address important unanswered questions. Consistent with previous ADQI meetings, a modified Delphi approach was followed. We performed a selected search and review of the available literature pre-conference, as described in detail elsewhere in this volume. We focused on the task of identifying various techniques to characterize the physiological changes that are encountered during the development and course of AKI and determine their utility as physiological biomarkers for patient management. Studies were identified via PubMed and Web of Science using the term 'biomarker', and acute kidney injury '(AKI)' combined with 'physiology' and 'functional'. The large body of literature retrieved generated a series of key questions, which were used to limit the scope of the review. Only representative publications are cited in this review.

Results

The ADQI breakout groups identified important new clinical needs in AKI including: (1) Do physiological biomarkers improve the risk assessment of AKI? If so, what set or group of physiological biomarkers is essential for assessment of the functional state of the kidney? (2) Do real-time physiological biomarkers improve the early diagnosis and aid in the differential diagnosis of AKI? (3) What is the interplay or added value of using physiological biomarkers along with biochemical markers of AKI and clinical characteristics in the diagnostic and prognostic assessment of AKI? (4) Can real-time monitoring of physiological biomarkers be used to guide therapy? These questions cannot be answered at this time, but it is our strong belief that in the next 5 years, we will begin to answer some of these important questions.

In this brief review, we summarize the current-state-of-the-art of available techniques to monitor physiological biomarkers, and propose a road map aimed at providing a reliable and quantifiable platform for the measurement of renal physiological biomarkers. Since this is an emerging area, techniques are currently limited and under development.

Renal Blood Flow, Tissue Oxygenation and Microcirculatory Function: Critical Physiological Biomarkers

Adequate parenchymal oxygen is an essential physiological requirement for sustaining kidney cell viability. In the kidney, 90% of oxygen consumption sustains oxidative phosphorylation for ATP production and sodium reabsorption [2]. Since sodium reabsorption is intrinsically dependent on sodium load, renal oxygen consumption is in turn highly dependent on the GFR and renal blood flow (RBF) [3]. RBF, blood flow distribution between cortex and medulla and microcirculatory blood flow may change prior to changes in damage biomarkers and may serve as one of the earliest indicators of altered renal function and complement damage biomarkers. Hyperfiltration and an increase in RBF [4] may reflect differences in 'renal reserve' and assessment of this parameter may provide information on risk of AKI. Autoregulation maintains normal GFR and RBF when mean arterial pressures drop to ~80 mm Hg. This is accomplished through a decrease in afferent arteriolar tone mediated by vasodilators such as prostaglandins and increase in efferent arteriolar tone mediated by angiotensin II. Prior to tubular injury, alterations in RBF may represent the earliest measurable physiological changes.

In the absence of adequate oxygen supply, the kidney is at risk for failure [5]. Pathological processes such as AKI disrupt the balance between tissue oxygen, nitric oxide and radical oxygen species [6].

Table 1. Physiological biomarkers for assessment of kidney function in AKI

Glomerular filtration rate/urine flow monitoring
– Urine indices
– Real-time GFR measurement
– Serial serum creatinine measurement (with correction for fluid balance using bioelectrical impedance analysis)
– Continuous urine flow
Renal perfusion
– Doppler ultrasound (visualization of macrocirculation)
– Contrast-enhanced ultrasound (visualization of microcirculation)
Renal oxygenation
– Bladder tissue pO_2
– Bladder urine pO_2 (measure of renal medullary oxygenation)
– Near infrared spectroscopy (measure of renal O_2 bioavailability)
– BOLD MRI (measure of renal O_2 bioavailability)
– PET (measure of renal metabolism)
Other complimentary markers
– Kidney ultrasound
– Renal venous O_2 saturation (measure of renal oxygen consumption)
– Urinalysis (renal indices, urine sediment, flow cytometry)
– Endothelial markers (e.g. endothelial microparticles, glycocalyx degradation)
– Inflammatory markers (e.g. cytokines, immune cells)
– Oxidative stress markers

Existing and developing non-invasive techniques to measure physiological parameters of kidney function in AKI are listed in table 1. The known clinical characteristics of each of these techniques are included in table 2. For example, contrast-enhanced ultrasound (CEU) may be used to determine 'renal reserve' by measuring changes in RBF in response to a protein load [7]. This maneuver could potentially identify subjects who lack reserve capacity and may be at higher risk for AKI. Near infrared spectroscopy or blood oxygenation level-dependent (BOLD) magnetic resonance imaging (MRI) may be used for detection of changes in oxygenation of the kidney. Detection of changes in oxygenation could identify those individuals at risk for AKI.

Glomerular Filtration Rate, Continuous Urine Output Measurement, and Urine Indices

Real-Time GFR Measurement

Determination of GFR is of utmost importance for evaluating kidney function in patients with kidney disease and for appropriate drug dosing. The clearance of an ideal filtration marker such as inulin, iothalamate or iohexol, is the gold standard

Table 2. Characteristics of physiological biomarkers of AKI

Physiological biomarker	Parameter	Unit	Bed-side	Inva-sive	Con-tinuous	Clinical perfor-mance	Cost
Urine indices	Urine Na, Fe_{Na}, Fe_{urea} osmolality	mEq/l, AU, mosm/kg H_2O	+	–	–	+	+
Serial serum creatinine	Estimated GFR	mg/dl	+	–	–	?	+
Real-time measured GFR	Measured GFR	ml/min	+	+	–	?	+++
Continuous urine flow	Urine output	ml/kg/h	+	–	+	?	+
Doppler ultrasound	Macrocirculation	resistive index	+	–	–	?	++
Contrast-enhanced ultrasound	Macro/microcirculation	AU	+	+	–	?	+++
Urine pO_2	Renal medullary O_2 tension	mm Hg	+	–	+	?	++
Bladder pO_2	Tissue O_2 tension	mm Hg	+	+	+	?	++
BOLD MRI	Renal O_2 availability	Hb O_2/Hb	–	–	–	?	++++
Positron emission tomography	O_2 uptake/renal metabolism	mCi/μg	–	+	–	?	++++
Near infrared spectroscopy	Renal O_2 availability	Hb O_2/Hb	+	–	+	?	++++
Bioelectrical impedance analysis	Renal tissue perfusion	AU	+	–	+	?	+

for measuring GFR [8]. However, these cumbersome, time-consuming techniques are not practical for use in clinical practice. A transcutaneously remote real-time analysis of FITC-sinistrin excretion has been used to determine GFR in healthy rats and rats with kidney disease [9]. In these studies, comparative measurements of transcutaneous and plasma elimination kinetics of FITC-sinistrin were compared, and highly comparable GFR values were demonstrated in all animal groups.

Lately, a point-of-care bedside fluorescence-based measured GFR (mGFR) assay has been developed experimentally for the rapid measurement of GFR [9, 10], which has potential applicability in AKI. In the main descriptive study, fluorescent conjugates of a small freely filterable reporter (inulin) and a large non-filterable marker (500 kDa dextran) were infused as a bolus, and the in vivo fluorescent signals were detected and quantified by the ratiometric optical fiber system [10]. The plasma volume was estimated by dilution of the large dextran molecule and the GFR was determined using the ratiometric two-compartment method 60 min after the injection. This method presented excellent agreement with the concurrent 6-hour iohexol-based GFR measurement technique.

In summary, fully developed real-time transcutaneous or intravascular monitoring of GFR may prove to be an ideal bedside tool for risk stratification, early diagnosis, prognosis and therapy guidance in AKI.

Serial Creatinine Measurement

Serum creatinine is an inexpensive and readily available physiological biomarker of kidney function and remains the 'gold standard' test for diagnosing AKI [11, 12], however it is non-specific for the nature and extent of injury and is influenced by several factors including the volume of distribution requiring adjustments for the level of fluid accumulation (prior to achievement of steady-state conditions) [13]. For these reasons, it is considered a suboptimal GFR marker [14] and a late biomarker of AKI when both damage and physiological changes occur together. More frequent measurements might allow earlier detection of AKI and could provide valuable insight into the mechanism and pathways to track the course of AKI. This hypothesis needs to be formally tested.

Continuous Urine Output Measurement

The importance of urine output in the detection of AKI [11] has recently been preliminarily verified [15], providing support for continuously monitoring urine output. A recent study assessed the use of an electronic urine output monitoring device in 20 critically ill patients [16]. Urine volume was measured using a drip detector, based on infrared detection. In this study, hourly urine output was measured with both traditional urinometer and continuous urine output monitoring device; both validated by cylinder measurements. Positive predictive value for AKI of a urine output of 40 ml/h was 91% with the continuous urine output monitoring device, and 77% with the urinometer. Additional studies are needed to further validate the performance of this user-friendly system and to prospectively evaluate whether urine output combined with serial creatinine measurements provide reliable, robust predictive tools in the setting of AKI.

Fractional Excretion of Sodium and Urea, and Urine Osmolality

Urine sodium and osmolality have been standard urine indices used in differentiating pre-renal AKI from acute tubular necrosis for decades. In a small study of patients with various forms of AKI, the fractional excretion of sodium (Fe_{Na}) proved to be useful. Although Fe_{Na} correctly diagnosed 86 of 87 patients, and urinary osmolality and urinary sodium correctly diagnosed AKI in 46 and 60% of patients, respectively [17], numerous exceptions limit the value of these clinical tests [18].

The measurement of the fractional excretion of urea (Fe_{urea}) may be useful in patients with AKI. Studies have documented an Fe_{urea} of 50–65% in patients with acute tubular necrosis compared to values below 35% in patients with pre-renal AKI. Although this urine index may be more useful than Fe_{urea} in patients who have received diuretics, it should not be used in isolation [19]. The traditional use of Fe_{Na} and Fe_{urea} will need to be reconsidered given the growing concept that pure functional loss may also be associated with damage and altered tubular function. For example, we will need to understand the diagnostic and prognostic significance of Fe_{Na} and Fe_{urea} of <1% and 35%, respectively, when damage biomarkers are positive.

Assessment of Renal Tissue Perfusion

Doppler Ultrasound

Using 2D ultrasonography and color Doppler, systolic and diastolic velocities of interlobar arteries can be measured. The Doppler resistive index (RI) described by Pourcelot ((peak systolic velocity – end diastolic velocity)/peak systolic velocity) is the traditional index used as a measure of renal vascular resistance. A value of 0.60 is considered to be a normal value for renal RI, whereas 0.70 is usually considered to be the upper threshold of normal RI in adults [20]. This index has been proposed for the quantification of the changes in renal vascular resistance but may be affected by other factors such as central hemodynamics (especially arterial stiffness, pulse pressure and heart rate), interstitial renal pressure, intra-abdominal pressure and other factors including age and underlying kidney disease. The RI has been found to reflect the degree of parenchymal damage in chronic kidney disease [21, 22] or hypertension [23], predict long-term kidney function in transplanted patients [24] and the diagnosis of obstructive uropathy [25].

Renal Doppler ultrasound has been used to predict AKI [26, 27]. In 37 critically patients with septic shock, the RI predicted the development of AKI 5 days after presentation (RI of 0.77 vs. 0.68 in AKI vs. non-AKI, respectively) [26]. These findings were recently confirmed in mechanically ventilated patients [27]. The RI was found to be higher in patients with acute tubular necrosis (0.85) than in pre-renal AKI (0.67), and better than urinary indices in the diagnosis of persistent AKI [28]. Finally, the RI might be useful in the assessment of renal circulation during the initial resuscitation in sepsis [29]. These studies indicate that Doppler is a simple, non-invasive tool that may be of significant utility in the setting of AKI (table 3).

Contrast-Enhanced Ultrasound

The gold standard measurement of RBF is *p*-aminohippurate (PAH) clearance. Although relatively accurate, this method has the disadvantage of being time-consuming and expensive. Furthermore, decreased renal tubular extraction in AKI and other high-risk settings (cardiopulmonary bypass, critical illness) largely invalidates this method of RBF measurement in such patients.

Contrast-enhanced ultrasonography, although relatively new in assessing renal perfusion, has been used extensively to assess myocardial perfusion during echocardiography. Contrast agents used for ultrasonography are gas-filled microbubbles (1–6 μm in diameter) which scatter ultrasound very effectively and are coated with a protein, lipid or polymer layer [30]. Second-generation microbubbles use inert, poorly soluble perfluorinated gases stabilized by phospholip-

Table 3. Utility of physiological biomarkers in AKI

Physiological biomarker	Risk assessment	Early diagnosis	Differential diagnosis	Prognosis	Therapy guidance
Urine indices	–	√	√	–	√
Serial serum creatinine	√	√	√	√	√
Real-time measured GFR	√	√	√	√	√
Continuous urine flow	√	√	√	√	?
Doppler ultrasound	√	√	√	?	√
Contrast-enhanced ultrasound	√	√	√	?	√
Urine pO_2	√	√	?	?	√
Bladder pO_2	√	√	√	?	√
BOLD MRI	?	?	√	√	?
Positron emission tomography	?	?	√	√	?
Near infrared spectroscopy	?	?	√	√	?
Bioelectrical impedance analysis	–	√	–	√	√

ids or albumin [30]. The small size allows microbubbles to provide information on small capillary beds. Although CEU has been used safely in humans, there are few studies on its use in AKI. In healthy volunteers, use of drugs to block the renin-angiotensin system produced measurable changes in microbubble perfusion indices that correlated with PAH clearance [31]. CEU has been used to diagnose renal artery stenosis, and was found to be more sensitive and specific in comparison to color Doppler ultrasound and similar to angiography [32]. In healthy volunteers, CEU detected a 43% increase in RBF following a high protein meal [7]. The ability to determine renal reserve could be useful in identifying patients who are at risk for AKI. Lastly, in kidney transplantation, CEU has been used to identify perfusion abnormalities helpful in the diagnosis of acute rejection [33] and chronic allograft nephropathy [34].

Bioelectrical Impedance Analysis (BIA)

BIA is a safe non-invasive, rapid and inexpensive technique to measure fluid compartment volume. It measures electrical properties of the body by measuring resistance, reactance and phase angle [35]. This method has been used in assessment nutritional status, urea distribution or dry weight during dialysis. It also has been used to assess intradialytic hypotension, as well as in AKI during CRRT. The use of BIA with serum creatinine is supported by recent data suggesting that further accuracy can be gained if serum creatinine is corrected for fluid accumulation [13]. The current method to assess fluid accumulation is through daily assessment of fluid balance. The use of BIA can give accurate assessment of total body water within minutes and can provide rapid correction of serum creatinine.

Assessment of Kidney Oxygenation and Metabolism

Positron Emission Tomography (PET)

PET is an imaging technique that produces three-dimensional images of physiological activity in the body, by detection of pairs of gamma rays emitted by a positron-emitting radionuclide tracer. The tracer is administered on a biologically active molecule, which tracks to specific areas of physiological activity and quantifies processes of biological importance, in an anatomical context. In the context of the kidney, PET scanning provides the ability to non-invasively quantify both RBF and kidney metabolism. PET measurement of RBF using ^{15}O-labeled water ($H_2{}^{15}O$) has been validated using radiolabelled microspheres in pigs [36]. Similar analyses in humans with and without kidney disease demonstrate that PET with $H_2{}^{15}O$ provides a non-invasive method for quantitative mapping of RBF [37]. PET imaging has also been used to evaluate kidney metabolism. In one such study, PET with ^{18}F-fluorodeoxyglucose was used to assess acute rejection in a rat renal transplantation model [38]. Further prospective studies are needed to determine the utility of PET scanning as a tool in the setting of AKI.

Optical Spectroscopy

Optical spectroscopy provides a non-invasive method of determining the oxygenation of blood in tissues. Optical spectroscopy uses differing optical absorbance of light in the near infrared (NIR) region (650–900 nm) between oxy- and deoxyhemoglobin to assess the oxygenation status in various vascular compartments. NIR spectroscopy devices entail an emitting light source directing light into tissues, and one or more sensors placed cutaneously measuring the returned light at several wavelengths. Currently, NIR spectroscopy devices for monitoring the kidney have only been used in children due to the relative accessibility of the pediatric kidney to NIR light in comparison to that in adults. NIR spectroscopy was first used to non-invasively monitor kidney oxygenation transcutaneously in mechanically ventilated children [39]. This study was able to monitor the effects of spontaneous episodes of oxygen desaturation on the kidney and determine the renal oxygen extraction during these episodes. Another study of NIR spectroscopy in children with heart failure was able to demonstrate a strong correlation between renal NIR spectroscopy values measured at the flank and renal vein saturation [40]. NIR spectroscopy has been used to demonstrate the ability of blood transfusions to effectively oxygenate the kidney in anemic children [41]. This use is of special importance due to the sensitivity of the kidney to hypoxemia and anemia [5]. A study in children undergoing cardiac surgery demonstrated a strong correlation between the low kidney oxygen saturation level and the development of AKI [42].

Direct access to the kidney could provide more direct and detailed measurements. Direct observation of the human renal microcirculation for this purpose has been accomplished by the use of hand-held microscopes [43]. Direct NIR spectroscopy measurement for assessment of renal cortex microcirculatory oxygenation has been combined with a miniature laser Doppler device to measure microcirculatory oxygenation and perfusion of the kidney cortex for assessing the success of transplantation [44]. Although the current applications are still sparse, it is expected that technical improvements will permit a comprehensive assessment of renal perfusion and oxygenation and provide a robust clinical platform to monitor the progress, response to therapy and resolution of AKI at the bedside.

BOLD MRI

BOLD MRI is a rapid non-invasive method for assessing renal tissue oxygen bioavailability, relying on the paramagnetic properties of deoxyhemoglobin [45]. The concentration of deoxyhemoglobin increases with rising oxygen consumption, leading to a decreasing T2 relaxation time of the surrounding tissue. Increased R2 (the rate of spin dephasing; equal to 1/T2 relaxation time) levels therefore imply increased deoxyhemoglobin (decreased oxyhemoglobin) and decreased partial pressures of oxygen (PaO_2) in renal tissues [45]. The capability of BOLD MRI to evaluate parenchymal oxygen bioavailability in many renal settings has been reported both experimentally and clinically [46–50]. There are fewer studies utilizing BOLD MRI in AKI [51], in acute rejection compared to kidney transplant recipients with prerenal azotemia or acute tubular necrosis [52].

Urine Oxygen (O_2) Tension

O_2 tension is the local partial pressure of O_2 in the urine (PuO_2). The pO_2 of urine entering the calyces reflects oxygen tension of the renal medulla, and is determined by O_2 supply, renal medullary blood flow, and oxygen consumption and may be the most reliable quantitative index of renal medullary perfusion.

Real-time PuO_2 has been measured in anesthetized dogs using the polarographic method [53], which is costly and cumbersome. A newer more promising and less expensive dynamic fluorescence quenching method relies on the collision of an O_2 molecule with a fluorophore in its excited state, which leads to a non-radiative transfer of energy [53]. MRI has also been used to quantify PuO_2 [54].

In studies of anesthetized human volunteers [55], the PuO_2 is fairly constant under normal conditions. The PuO_2 decreases in states of volume depletion and increases following volume re-expansion. In hemorrhagic shock, there is a sharp decline in PuO_2. Bladder PuO_2 has also been measured in settings of crit-

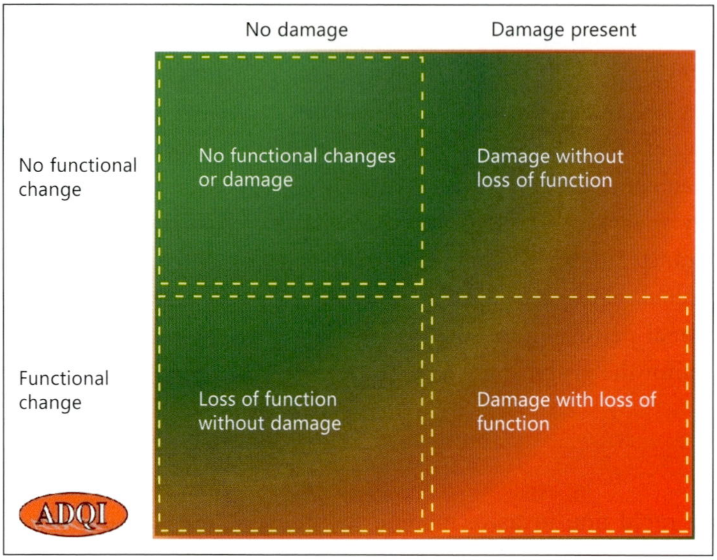

Fig. 2. Application of physiological biomarkers in the spectrum of AKI. Physiological biomarkers including imaging and other novel markers of kidney function can be used across the quadrants to establish normal or abnormal kidney physiological state. The choice of physiological biomarkers would be based on the contributing factors and correlated with clinical and laboratory findings. The physiological biomarkers would be utilized to determine transitions across these groups with sequential measurements. Which physiological biomarkers are used and the thresholds applied will need to be defined further in prospective studies. Reproduced with permission from ADQI [58].

ical illness [56], cardiac arrest [55], cardiogenic shock [55], and following on-pump cardiac surgery [57], and has been shown to improve by the use of mannitol [55] and fenoldopam [56]. In a study of 90 patients undergoing cardiopulmonary bypass surgery, the postoperative serum creatinine level was significantly higher in patients whose PuO_2 decreased after surgery, as compared with those whose PuO_2 remained constant or increased [57]. This study presents the possibility of using bladder PuO_2 monitoring for the early detection of postoperative AKI.

Application of Physiological Biomarkers for Assessment and Management of AKI

The emergence of novel markers of kidney function offers the promise to provide clinicians with a set of tools to efficiently and more accurately assess the underlying processes resulting from kidney injury and determine the course

Table 4. Research agenda on the potential role of physiological biomarkers in AKI

1. Physiological biomarkers are valid, testable markers of kidney function and they can provide normal kidney function variables during AKI
2. To test the hypothesis that physiological biomarkers are able to guide titration of fluid management
3. To test the hypothesis that physiological biomarkers are able to determine renal functional reserve
4. To test the hypothesis that physiological biomarkers improve precision of clinical and biochemical markers
5. To test the hypothesis that bioelectrical impedance analysis is valid for assessing volume status
6. To test the hypothesis that bioelectrical impedance analysis to correct for fluid overload in using serum creatinine improves precision over using serum creatinine alone

and time points for intervention. As illustrated in figure 2, we anticipate that the physiological biomarkers identified in this review will be combined with damage markers to classify patients into one of the four AKI categories at presentation, which could be used to track potential transitions from one category to another over time. The choice of physiological biomarkers for diagnosis and staging, differential diagnosis, and prognosis of AKI will need to be determined in prospective studies. We believe that an expanded tool kit of physiological biomarkers might significantly enhance our capabilities to recognize AKI early, and provide a better understanding of the pathophysiology of this heterogeneous condition.

Conclusions

New methods to assess physiological variables are needed to integrate with existing biomarkers for the assessment of AKI. Many of the concepts discussed in this review are supported by animal studies and early human studies. The challenge will be to identify a minimum set of essential physiological parameters related to renal function, integrate these with the measurement of chemical biomarkers measured at relevant intervals and identify therapeutic/preventive modalities in relation to the physiological state of the kidney monitored at the bedside. Table 4 lists a research agenda to guide investigation for the next 5 years. It is expected that the integration of these techniques into a comprehensive monitoring environment will provide the needed physiological characterization of the kidney together with chemical biomarkers and to monitor to characterize progress and resolution of AKI.

Acknowledgements

This work was the result of a two and half day meeting of the 10th Acute Dialysis Quality Initiative Consensus Conference held in Dublin, Ireland, August 30, 2011 to September 1, 2011.

References

1 Sutton TA, Fisher CJ, Molitoris BA: Microvascular endothelial injury and dysfunction during ischemic acute renal failure. Kidney Int 2002;62:1539–1549.
2 Welch WJ: Intrarenal oxygen and hypertension. Clin Exp Pharmacol Physiol 2006;33:1002–1005.
3 Evans RG, Gardiner BS, Smith DW, O'Connor PM: Intrarenal oxygenation: unique challenges and the biophysical basis of homeostasis. Am J Physiol Renal Physiology 2008;295:F1259–F1270.
4 Hostetter TH: Human renal response to meat meal. Am J Physiol 1986;250:F613–F618.
5 Legrand M, Mik EG, Balestra GM, Lutter R, Pirracchio R, Payen D, Ince C: Fluid resuscitation does not improve renal oxygenation during hemorrhagic shock in rats. Anesthesiology 2010;112:119–127.
6 Le Dorze M, Legrand M, Payen D, Ince C: The role of the microcirculation in acute kidney injury. Curr Opin Crit Care 2009;15: 503–508.
7 Kalantarinia K, Belcik JT, Patrie JT, Wei K: Real-time measurement of renal blood flow in healthy subjects using contrast-enhanced ultrasound. Am J Physiol Renal Physiol 2009; 297:F1129–F1134.
8 Dagher PC, Herget-Rosenthal S, Ruehm SG, Jo SK, Star RA, Agarwal R, Molitoris BA: Newly developed techniques to study and diagnose acute renal failure. J Am Soc Nephrol 2003;14:2188–2198.
9 Schock-Kusch D, Xie Q, Shulhevich Y, Hesser J, Stsepankou D, Sadick M, Koenig S, Hoecklin F, Pill J, Gretz N: Transcutaneous assessment of renal function in conscious rats with a device for measuring FITC-sinistrin disappearance curves. Kidney Int 2011;79: 1254–1258.
10 Wang E, Meier DJ, Sandoval RM, Von Hendy-Willson VE, Pressler BM, Bunch RM, Alloosh M, Sturek MS, Schwartz GJ, Molitoris BA: A portable fiberoptic ratiometric fluorescence analyzer provides rapid point-of-care determination of glomerular filtration rate in large animals. Kidney Int 2012;81: 112–117.
11 Bellomo R, Ronco C, Kellum JA, Mehta RL, Palevsky P: Acute renal failure – definition, outcome measures, animal models, fluid therapy and information technology needs: the Second International Consensus Conference of the Acute Dialysis Quality Initiative (ADQI) Group. Crit Care 2004; 8:R204–R212.
12 Mehta RL, Kellum JA, Shah SV, Molitoris BA, Ronco C, Warnock DG, Levin A, Acute Kidney Injury Network: Report of an initiative to improve outcomes in acute kidney injury. Crit Care 2007;2007:R31.
13 Macedo E, Bouchard J, Soroko SH, Chertow GM, Himmelfarb J, Ikizler TA, Paganini EP, Mehta RL: Fluid accumulation, recognition and staging of acute kidney injury in critically ill patients. Crit Care 2010;14:R82.
14 Stevens LA, Coresh J, Greene T, Levey AS: Assessing kidney function – measured and estimated glomerular filtration rate. N Engl J Med 2006;354:2473–2483.

15 Macedo E, Malhotra R, Claure-Del Granado R, Fedullo P, Mehta RL: Defining urine output criterion for acute kidney injury in critically ill patients. Nephrol Dial Transplant 2011;26:509–515.
16 Hersch M, Einav S, Izbicki G: Accuracy and ease of use of a novel electronic urine output monitoring device compared with standard manual urinometer in the intensive care unit. J Crit Care 2009;24:629 e613–e627.
17 Espinel CH, Gregory AW: Differential diagnosis of acute renal failure. Clin Nephrol 1980;13:73–77.
18 Zarich S, Fang LS, Diamond JR: Fractional excretion of sodium. Exceptions to its diagnostic value. Arch Intern Med 1985;145: 108–112.
19 Pepin MN, Bouchard J, Legault L, Ethier J: Diagnostic performance of fractional excretion of urea and fractional excretion of sodium in the evaluations of patients with acute kidney injury with or without diuretic treatment. Am J Kidney Dis 2007;50:566–573.
20 Keogan MT, Kliewer MA, Hertzberg BS, DeLong DM, Tupler RH, Carroll BA: Renal resistive indexes: variability in Doppler US measurement in a healthy population. Radiology 1996;199:165–169.
21 Splendiani G, Parolini C, Fortunato L, Sturniolo A, Costanzi S: Resistive index in chronic nephropathies: predictive value of renal outcome. Clin Nephrol 2002;57:45–50.
22 Sugiura T, Nakamori A, Wada A, Fukuhara Y: Evaluation of tubulointerstitial injury by Doppler ultrasonography in glomerular diseases. Clin Nephrol 2004;61:119–126.
23 Derchi LE, Leoncini G, Parodi D, Viazzi F, Martinoli C, Ratto E, Vettoretti S, Vaccaro V, Falqui V, Tomolillo C, Deferrari G, Pontremoli R: Mild renal dysfunction and renal vascular resistance in primary hypertension. Am J Hypertens 2005;18:966–971.
24 Akgul A, Ibis A, Sezer S, Basaran C, Usluogullari A, Ozdemir FN, Arat Z, Haberal M: Early assessment of renal resistance index and long-term renal function in renal transplant recipients. Ren Fail 2009;31:18–24.
25 Platt JF, Rubin JM, Ellis JH: Acute renal obstruction: evaluation with intrarenal duplex Doppler and conventional US. Radiology 1993;186:685–688.
26 Lerolle N, Guerot E, Faisy C, Bornstain C, Diehl JL, Fagon JY: Renal failure in septic shock: predictive value of Doppler-based renal arterial resistive index. Intensive Care Med 2006;32:1553–1559.
27 Darmon M, Schortgen F, Vargas F, Liazydi A, Schlemmer B, Brun-Buisson C, Brochard L: Diagnostic accuracy of Doppler renal resistive index for reversibility of acute kidney injury in critically ill patients. Intensive Care Med 2011;37:68–76.
28 Platt JF, Rubin JM, Ellis JH: Acute renal failure: possible role of duplex Doppler US in distinction between acute prerenal failure and acute tubular necrosis. Radiology 1991; 179:419–423.
29 Deruddre S, Cheisson G, Mazoit JX, Vicaut E, Benhamou D, Duranteau J: Renal arterial resistance in septic shock: effects of increasing mean arterial pressure with norepinephrine on the renal resistive index assessed with Doppler ultrasonography. Intensive Care Med 2007;33:1557–1562.
30 Klibanov AL: Ultrasound contrast agents: development of the field and current status. Top Curr Chem 2002;222:73–106.
31 Kishimoto N, Mori Y, Nishiue T, Nose A, Kijima Y, Tokoro T, Yamahara H, Okigaki M, Kosaki A, Iwasaka T: Ultrasound evaluation of valsartan therapy for renal cortical perfusion. Hypertens Res 2004;27: 345–349.
32 Ciccone MM, Cortese F, Fiorella A, Scicchitano P, Cito F, Quistelli G, Pertosa G, D'Agostino R, Guida P, Favale S: The clinical role of contrast-enhanced ultrasound in the evaluation of renal artery stenosis and diagnostic superiority as compared to traditional echo-color-Doppler flow imaging. Int Angiol 2011;30:135–139.
33 Benozzi L, Cappelli G, Granito M, Davoli D, Favali D, Montecchi MG, Grossi A, Torricelli P, Albertazzi A: Contrast-enhanced sonography in early kidney graft dysfunction. Transplant Proc 2009;41:1214–1215.
34 Schwenger V, Korosoglou G, Hinkel UP, Morath C, Hansen A, Sommerer C, Dikow R, Hardt S, Schmidt J, Kucherer H, Katus HA, Zeier M: Real-time contrast-enhanced sonography of renal transplant recipients predicts chronic allograft nephropathy. Am J Transplant 2006;6:609–615.

35 Di Iorio BR, Scalfi L, Terracciano V, Bellizzi V: A systematic evaluation of bioelectrical impedance measurement after hemodialysis session. Kidney Int 2004;65:2435–2440.
36 Juillard L, Janier MF, Fouque D, Lionnet M, Le Bars D, Cinotti L, Barthez P, Gharib C, Laville M: Renal blood flow measurement by positron emission tomography using ^{15}O-labeled water. Kidney Int 2000;57:2511–2518.
37 Alpert NM, Rabito CA, Correia DJ, Babich JW, Littman BH, Tompkins RG, Rubin NT, Rubin RH, Fischman AJ: Mapping of local renal blood flow with PET and $H_2{}^{15}O$. J Nucl Med 2002;43:470–475.
38 Reuter S, Schnockel U, Schroter R, Schober O, Pavenstadt H, Schafers M, Gabriels G, Schlatter E: Non-invasive imaging of acute renal allograft rejection in rats using small animal F-FDG-PET. PLoS One 2009; 4:e5296.
39 Petrova A, Mehta R: Near-infrared spectroscopy in the detection of regional tissue oxygenation during hypoxic events in preterm infants undergoing critical care. Pediatr Crit Care Med 2006;7:449–454.
40 Ortmann LA, Fontenot EE, Seib PM, Eble BK, Brown R, Bhutta AT: Use of near-infrared spectroscopy for estimation of renal oxygenation in children with heart disease. Pediatr Cardiol 2011;32:748–753.
41 Dani C, Pratesi S, Fontanelli G, Barp J, Bertini G: Blood transfusions increase cerebral, splanchnic, and renal oxygenation in anemic preterm infants. Transfusion 2010;50:1220–1226.
42 Owens GE, King K, Gurney JG, Charpie JR: Low renal oximetry correlates with acute kidney injury after infant cardiac surgery. Pediatr Cardiol 2011;32:183–188.
43 Yamamoto T, Hayashi K, Matsuda H, Tomura Y, Ogasawara Y, Hashimoto R, Tada T, Tanaka H, Kajiya F: Direct in vivo visualization of glomerular microcirculation by intravital pencil lens-probe CCD videomicroscopy. Clin Hemorheol Microcirc 2000;23:103–108.
44 Scheeren TW, Martin K, Maruschke M, Hakenberg OW: Prognostic value of intraoperative renal tissue oxygenation measurement on early renal transplant function. Transpl Int 2011;24:687–696.
45 Prasad PV, Chen Q, Goldfarb JW, Epstein FH, Edelman RR: Breath-hold R2* mapping with a multiple gradient-recalled echo sequence: application to the evaluation of intrarenal oxygenation. J Magn Res Imaging 1997; 7:1163–1165.
46 Djamali A, Sadowski EA, Muehrer RJ, Reese S, Smavatkul C, Vidyasagar A, Fain SB, Lipscomb RC, Hullett DH, Samaniego-Picota M, Grist TM, Becker BN: BOLD-MRI assessment of intrarenal oxygenation and oxidative stress in patients with chronic kidney allograft dysfunction. Am J Physiol Renal Physiol 2007;292:F513–F522.
47 Wang ZJ, Kumar R, Banerjee S, Hsu CY: Blood oxygen level-dependent (BOLD) MRI of diabetic nephropathy: preliminary experience. J Magn Res Imaging 2011;33:655–660.
48 Thoeny HC, Zumstein D, Simon-Zoula S, Eisenberger U, De Keyzer F, Hofmann L, Vock P, Boesch C, Frey FJ, Vermathen P: Functional evaluation of transplanted kidneys with diffusion-weighted and BOLD MR imaging: initial experience. Radiology 2006; 241:812–821.
49 Textor SC, Glockner JF, Lerman LO, Misra S, McKusick MA, Riederer SJ, Grande JP, Gomez SI, Romero JC: The use of magnetic resonance to evaluate tissue oxygenation in renal artery stenosis. J Am Soc Nephrol 2008; 19:780–788.
50 Pedersen M, Dissing TH, Morkenborg J, Stodkilde-Jorgensen H, Hansen LH, Pedersen LB, Grenier N, Frokiaer J: Validation of quantitative BOLD MRI measurements in kidney: application to unilateral ureteral obstruction. Kidney Int 2005;67:2305–2312.
51 Juillard L, Lerman LO, Kruger DG, Haas JA, Rucker BC, Polzin JA, Riederer SJ, Romero JC: Blood oxygen level-dependent measurement of acute intra-renal ischemia. Kidney Int 2004;65:944–950.
52 Han F, Xiao W, Xu Y, Wu J, Wang Q, Wang H, Zhang M, Chen J: The significance of BOLD MRI in differentiation between renal transplant rejection and acute tubular necrosis. Nephrol Dial Transplant 2008;23: 2666–2672.
53 Shaw AD, Li Z, Thomas Z, Stevens CW: Assessment of tissue oxygen tension: comparison of dynamic fluorescence quenching and polarographic electrode technique. Crit Care 2002;6:76–80.

54 Wang Z, Joe B, Coakley F, Zaharchuk G, Busse R, Yeh B: Urinary oxygen tension measurement in humans using magnetic resonance imaging. Acad Radiol 2008;15: 1467–1473.
55 Leonhardt KO, Landes RR: Oxygen tension of the urine and renal structures. Preliminary report of clinical findings. N Engl J Med 1963;269:115–121.
56 Morelli A, Rocco M, Conti G, Orecchioni A, Alberto De Blasi R, Coluzzi F, Pietropaoli P: Monitoring renal oxygen supply in critically-ill patients using urinary oxygen tension. Anesth Analg 2003;97:1764–1768.
57 Kainuma M, Yamada M, Miyake T: Continuous urine oxygen tension monitoring in patients undergoing cardiac surgery. J Cardiothorac Vasc Anesth 1996;10:603–608.
58 Acute Dialysis Quality Initiative (ADQI). www.ADQI.org (accessed January 10, 2013).

Mark D. Okusa, MD
Division of Nephrology
University of Virginia Health System
Box 800133, Charlottesville, VA, 22908 (USA)
E-Mail mdo7y@virginia.edu

McCullough PA, Kellum JA, Mehta RL, Murray PT, Ronco C (eds): ADQI Consensus on AKI Biomarkers and Cardiorenal Syndromes. Contrib Nephrol. Basel, Karger, 2013, vol 182, pp 82–98
DOI: 10.1159/000349966

Pathophysiology of the Cardiorenal Syndromes: Executive Summary from the Eleventh Consensus Conference of the Acute Dialysis Quality Initiative (ADQI)

Peter A. McCullough[a] · John A. Kellum[b] · Michael Haase[c] · Christian Müller[d] · Kevin Damman[e] · Patrick T. Murray[f] · Dinna Cruz[g] · Andrew A. House[h] · Kai M. Schmidt-Ott[i] · Giorgio Vescovo[j] · Sean M. Bagshaw[k] · Eric A. Hoste[l] · Carlos Briguori[m] · Branko Braam[n] · Lakhmir S. Chawla[o] · Maria R. Costanzo[p] · James A. Tumlin[q] · Charles A. Herzog[r] · Ravindra L. Mehta[s] · Hamid Rabb[t] · Andrew D. Shaw[u] · Kai Singbartl[v] · Claudio Ronco[w] · for the Acute Dialysis Quality Initiative (ADQI) Consensus Group

[a]St. John Providence Health System, Warren, Mich., Providence Hospitals and Medical Centers, Southfield and Novi, Mich., St. John Macomb Oakland Center, Madison Heights, Mich., St. John Hospital and Medical Center, Detroit, Mich., USA; [b]Clinical Research, Investigation, and Systems Modeling of Acute Illness Center, Department of Critical Care Medicine, University of Pittsburgh, Pittsburgh, Pa., USA; [c]Department of Nephrology, Hypertension, Diabetes and Endocrinology, Otto von Guericke University, Magdeburg, Germany; [d]University Hospital, Basel, Switzerland; [e]Department of Cardiology, University Medical Center Groningen, Groningen, The Netherlands; [f]School of Medicine and Medical Science, University College Dublin, Dublin, Ireland; [g]Department of Nephrology Dialysis and Transplantation, San Bortolo Hospital, Vicenza, Italy; [h]Schulich School of Medicine and Dentistry, University of Western Ontario Division of Nephrology, University Hospital, London, Ont., Canada; [i]Experimental and Clinical Research Center, Charité-Universitätsmedizin Berlin, Max Delbrück Center for Molecular Medicine, Berlin, Germany; [j]Department of Internal Medicine, S. Bortolo Hospital, Vicenza, Italy; [k]Division of Critical Care Medicine, Faculty of Medicine and Dentistry, University of Alberta, Edmonton, Alta., Canada; [l]Department of Intensive Care Medicine, Ghent University Hospital, Gent, Belgium; [m]Department of Cardiology, Clinica Mediterranea, Naples, Italy; [n]Department of Medicine/Division of Nephrology and Immunology, University of Alberta Hospital, Edmonton, Alta., Canada; [o]Department of Anesthesiology and Critical Care Medicine and Division of Renal Diseases and Hypertension, Department of Medicine, George Washington University Medical Center, Washington, D.C., USA; [p]Midwest Heart Foundation, Edward Heart Hospital, Naperville, Ill., USA; [q]Department of Medicine, College of Medicine, University of Tennessee, Chattanooga, Tenn., USA; [r]Hennepin County Medical Center, Minneapolis,

ADQI 11 Workgroup members are listed in Appendix 2.

Minn., USA; [s]Division of Nephrology and Hypertension, Department of Medicine, University of California San Diego, San Diego, Calif., USA; [t]Nephrology Division, Johns Hopkins Hospital and Department of Medicine, Johns Hopkins University School of Medicine, Baltimore, Md., USA; [u]Department of Anesthesiology, Duke University Medical Center, Durham, N.C., USA; [v]Department of Critical Care Medicine, Clinical Research, Investigation, and Systems Modeling of Acute Illness Center, University of Pittsburgh, Pittsburgh, Pa., USA; [w]Department of Nephrology, International Renal Research Institute (IRRIV), San Bortolo Hospital, Vicenza, Italy

Abstract

Cardiorenal syndromes (CRS) have been recently classified into five distinct entities, each with different major pathophysiologic mechanisms. CRS type 1 most commonly occurs in the setting of acutely decompensated heart failure where approximately 25% of patients develop a rise in serum creatinine and a reduction of urine output after the first several doses of intravenous diuretics. Altered cardiac and renal hemodynamics are believed to be the most important determinants of CRS type 1. CRS type 2 is the hastened progression of chronic kidney disease (CKD) in the setting of chronic heart failure. Accelerated renal cell apoptosis and replacement fibrosis is considered to be the dominant mechanism. CRS type 3 is acutely decompensated heart failure after acute kidney injury from inflammatory, toxic, or ischemic insults. This syndrome is precipitated by salt and water overload, acute uremic myocyte dysfunction, and neurohormonal dysregulation. CRS type 4 is manifested by the acceleration of the progression of chronic heart failure in the setting of CKD. Cardiac myocyte dysfunction and fibrosis, so-called 'CKD cardiomyopathy', is believed to be the predominant pathophysiologic mechanism. Type 5 CRS is simultaneous acute cardiac and renal injury in the setting of an overwhelming systemic insult such as sepsis. In this scenario, the predominant pathophysiological disturbance is microcirculatory dysfunction as a result of acutely abnormal immune cell signaling, catecholamine cellular toxicity, and enzymatic activation which result in simultaneous organ injury often extending beyond both the heart and the kidneys. This paper will summarize these and other key findings from an international consensus conference on the spectrum of pathophysiologic mechanisms at work in the CRS.

The term cardiorenal syndromes (CRS) was coined by Ledoux [1] in 1951 and referred to combined heart and kidney failure. In 1997 to the present, Schrier and colleagues [2–4] in multiple papers summarized the impact of salt and water retention combined with neurohormonal activation in the pathogenesis of CRS. Brammah et al. [5] pointed out that treatment of bilateral renal arterial disease could result in improvement of both heart and kidney function. Braam and co-workers [6] in 2005 proposed that organ dysfunction in one system influences that in the other. In 2008, Ronco et al. [7] proposed five distinct CRS according to the temporal sequence of organ injury and failure as well as the clinical context. Later in 2008, a consensus conference was organized under the auspices of the Acute Dialysis Quality Initiative (ADQI) bringing to Venice, Italy, key opinion leaders and experts in the fields of nephrology, critical care, cardiac surgery, and cardiology. In series of papers and an executive summary, this group presented the definition and classification system for the CRS, epidemiology and diagnostic criteria, prevention and therapeutic approaches, as a result of the consensus process [8]. In 2012, ADQI conducted a second consensus conference to review and summarize the rapidly developing literature on the spectrum of pathophysiologic mechanisms involved in CRS.

Methods

The ADQI process was applied using previously described methodology taking advantage of key opinion leaders in the field and identifying appropriate topics for consensus [9]. The ADQI methodology comprises a systematic search for evidence with review and evaluation of relevant literature, establishment of clinical and physiologic outcomes for comparison of different treatments, description of current practice and analysis of areas in which evidence is lacking and a specific research agenda is required. ADQI activities were divided into a pre-conference, conference, and post-conference phase. Before the conference, topics were selected and workgroups assembled. Groups identified key questions and conducted a systematic literature search. During the conference, workgroups assembled in breakout sessions, as well as plenary sessions where their findings were presented, debated, and refined. Key questions were identified by the entire ADQI group, and the subgroups deliberated on these questions, bringing forth recommendations to the group as a whole. Deliberations followed three days of discussion amongst 23 attendees. Summary statements were then developed by the entire group and reported into the present document.

Results

Informed by previous publication records and expertise, the Steering Committee assembled a panel representing multiple disciplines from a variety of countries and scientific societies. From this plenary group, five smaller working groups were identified to examine the predominant pathophysiologic mechanism of CRS types 1–5.

Principles of Definition and Classification

It was agreed in the first ADQI consensus that a large umbrella term be preferred, using the plural *(Cardiorenal Syndromes)*, to indicate the presence of multiple syndromes (fig. 1). Subtypes would recognize the primary organ dysfunction (cardiac versus renal) as well as the acute versus chronic nature of the condition (fig. 2). Both organs must have or develop structural or func-

Cardiorenal Syndrome (CRS) General Definition: *A pathophysiologic disorder of the heart and kidneys whereby acute or chronic dysfunction in one organ may induce acute or chronic dysfunction in the other organ*
CRS Type I (Acute Cardiorenal Syndrome) *Abrupt worsening of cardiac function (e.g. Acutely decompensated congestive heart failure) leading to acute kidney injury*
CRS Type II (Chronic Cardiorenal Syndrome) *Chronic abnormalities in cardiac function (e.g. chronic congestive heart failure) causing progressive and permanent chronic kidney disease*
CRS Type III (Acute Renocardiac Syndrome) *Abrupt worsening of renal function (e.g. acute kidney injury) causing acute cardiac disorder (acute heart failure)*
CRS Type IV (Chronic Renocardiac Syndrome) *Chronic kidney disease (diabetic nephropathy) contributing to decreased cardiac function, cardiac hypertrophy, fibrosis, and/or increased risk of adverse cardiovascular events*
CRS Type V (Secondary Cardiorenal Syndrome) *Systemic condition (e.g. sepsis) causing both acute cardiac and renal injury and dysfunction*

Fig. 1. Five subtypes of CRS as proposed by Ronco and the AQDI.

tional abnormalities to fulfill the criteria for definition. An additional subtype was desired to capture systemic conditions that affect both organs simultaneously.

The umbrella term *'Cardiorenal Syndromes' (CRS)* was defined as 'Disorders of the heart and kidneys whereby acute or chronic dysfunction in one organ may induce acute or chronic dysfunction of the other'. Five subtypes of the syndromes were identified and defined as reported in figure 1. All subtypes imply that organ crosstalk is involved in the global acute and chronic pathogenesis (fig. 2) [10].

Acute Cardiorenal Syndrome (Type 1): Acute Worsening of Heart Function Leading to Kidney Injury and/or Dysfunction

This is a syndrome of worsening renal function that frequently complicates acutely decompensated heart failure (ADHF). Between 27 and 40% of patients hospitalized for ADHF develop acute kidney injury (AKI) as defined by an increase in serum creatinine of 0.3 mg/dl or more and fall into this clinical entity [11, 12]. These patients experience higher mortality and morbidity, and increased length of hospitalization.

ADHF can be phenotypically characterized in a schematic and two-by-two table as shown in figure 3 [13]. Here, the relative degree of systemic congestion and perfusion can be used to organize renal hemodynamic processes that are believed to be related to CRS type 1 [14]. Passive venous congestion influences

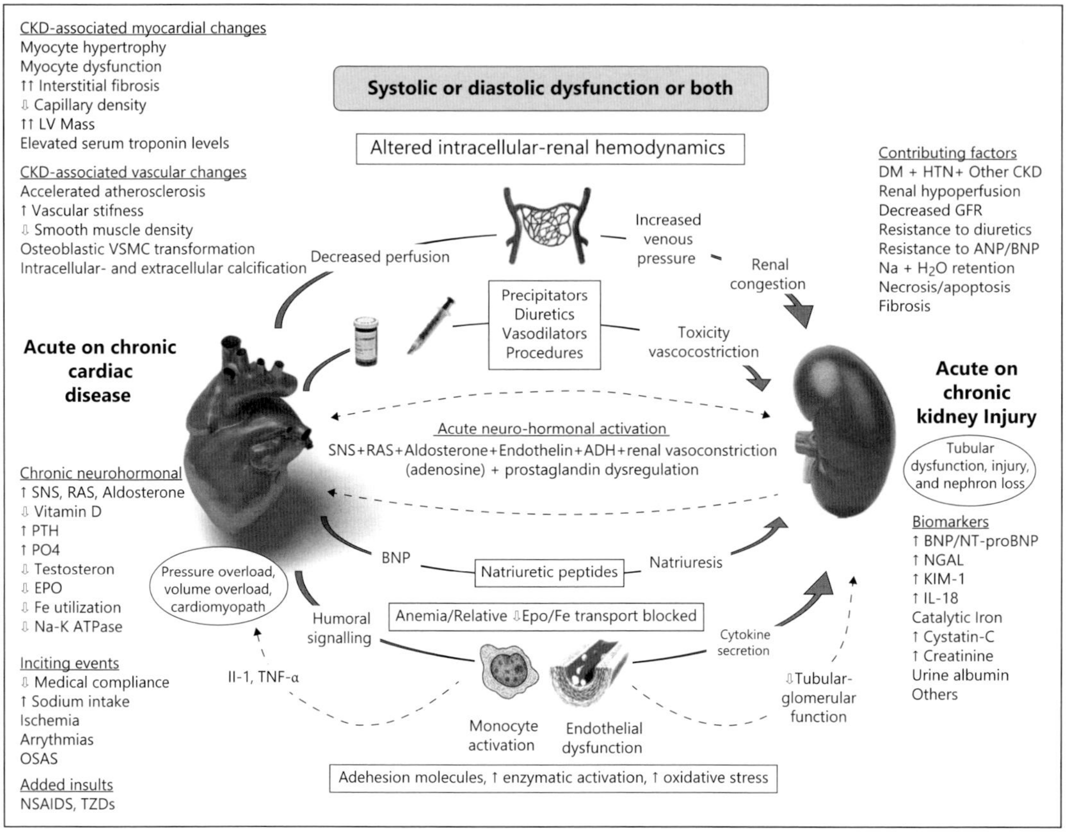

Fig. 2. Acute on chronic elements of disease observed in CRS.

the efferent flow from the glomerulus, the peritubular network, as well as the arcades of renal veins that coalescence the major renal vein and its accessory veins that carry blood to the inferior vena cava. Several studies have shown that renal venous congestion is an important risk predictor for the development of worsened renal filtration function in hospitalized patients [15]. Proposed pathophysiologic mechanisms include reduced transglomerular pressure, elevated renal interstitial pressure, myogenic and neural reflexes, baroreceptor stimulation, activation of sympathetic nervous and renin-angiotensin-aldosterone systems, non-osmotic release of arginine vasopressin, local production of endothelin, and enhanced proinflammatory pathways. When levels of sympathetic and renin-angiotensin system activity are further raised after one or two doses of intravenous loop diuretics, in patients with susceptible renal hemodynamics, the CRS type 1 can be initiated and recognized over the next several hours as a reduction in diuretic response and later a rise in serum creatinine [16]. Biomarker evidence of neurohormonal activation includes the natriuretic peptides, mid-

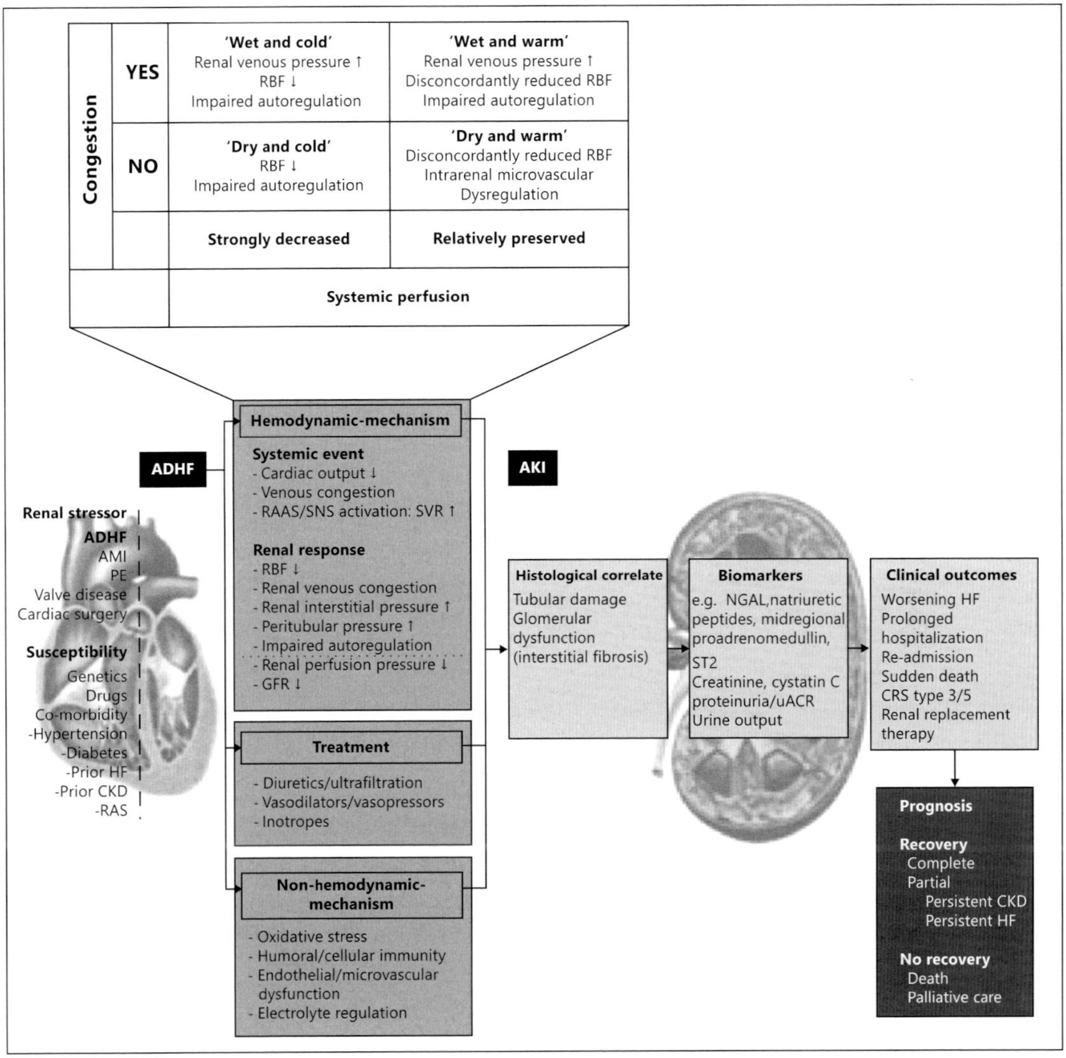

Fig. 3. Pathophysiology of CRS type 1.

regional proadrenomedullin, copeptin (C-terminal segment of preprovasopressin) and procalcitonin [17, 18]. In refractory patients, hyponatremia as a physiologic reflection of marked neurohormonal activation in CRS type 1 heralds a very poor prognosis [19]. Recent studies suggest individuals with greater levels of natriuretic peptides as baseline and evidence of pulmonary congestion, in addition to increased central venous pressure, have the greatest risk for CRS type 1 [20]. The end result of these hemodynamic perturbations of the kidney result in a loss of autoregulation and the onset of worsened salt and water retention, reduction in renal filtration, and oliguria.

Beyond hemodynamics and neurohormonal activation, abnormal patterns of cell signaling have been linked to ADHF and possibly CRS. Soluble ST-2 is a recently approved blood biomarker in the risk prediction of heart failure (HF) hospitalization and death [21, 22]. This marker appears to be complementary to the natriuretic peptides and reflects the degree of biomechanical strain and immune cell activation and cell signaling that occurs in progressive HF [23]. ST2 is the receptor for interleukin (IL)-33, a cytokine with angiohypertrophic and antifibrotic effects on the myocardium [24]. Soluble ST is a circulating inhibitor of the ST2 receptor, and hence works to negate the beneficial effect of IL-33. The impact on cell signaling, the proximal tubule, and sequential effects on tubuloglomerular feedback mediated by adenosine, nitric oxide, prostaglandins and other substances in the macula densa cells, are not fully understood at this time [25]. However, this line of reasoning could explain how systemic cell signaling could trigger events in the kidney in the form of maladaptive tubuloglomerular feedback which cause a transient reduction in renal filtration and a rise in serum creatinine [26, 27].

Chronic CRS (Type 2): Chronic Abnormalities in Heart Function Leading to Kidney Injury or Dysfunction

This subtype indicates a more chronic HF hastening the progression of chronic kidney disease (CKD). Approximately 63% of patients with HF meet the definition of stage 3–5 CKD with an estimated glomerular filtration rate <60 ml/min/1.73 m^2 [28]. There is considerable evidence to support the observation that HF is a proinflammatory state characterized by markedly elevated levels of tumor necrosis factor-α, upregulation of soluble receptors for tumor necrosis factor, and a number of ILs including IL-1β, IL-18 and IL-6, as well as several cellular adhesion molecules. Multiple studies have indicated that chronic HF induces microalbuminuria, probably through cytokine-induced damage to the endothelial glycocalyx, and thus, creates an additional burden on the proximal tubules (megalin-cubilin complex) in terms of protein reabsorption [29–31]. This process can be the focal point for intrarenal cell signaling and inflammation that leads to progressive functional and anatomic loss of nephrons [32]. Recently, HF has been shown to induce the distal renal tubular cells to upregulate the production of neutrophil gelatinase-associated lipocalin (NGAL), a reliable indicator of both chronic and acute kidney disease [33].

As a result of chronic volume overload and persistent hyperactivation of the neurohormonal (renin-angiotensin-aldosterone) and sympathetic nervous systems, functional and structural changes occur within the kidneys. Through paracrine activation, HF is associated with activation of macrophages and secretion of galectin-3. Galectin-3 is one of a family of lectins (otherwise known as MAC-2 Ag) whose main target is myofibroblasts via intracellular signaling, which trans-

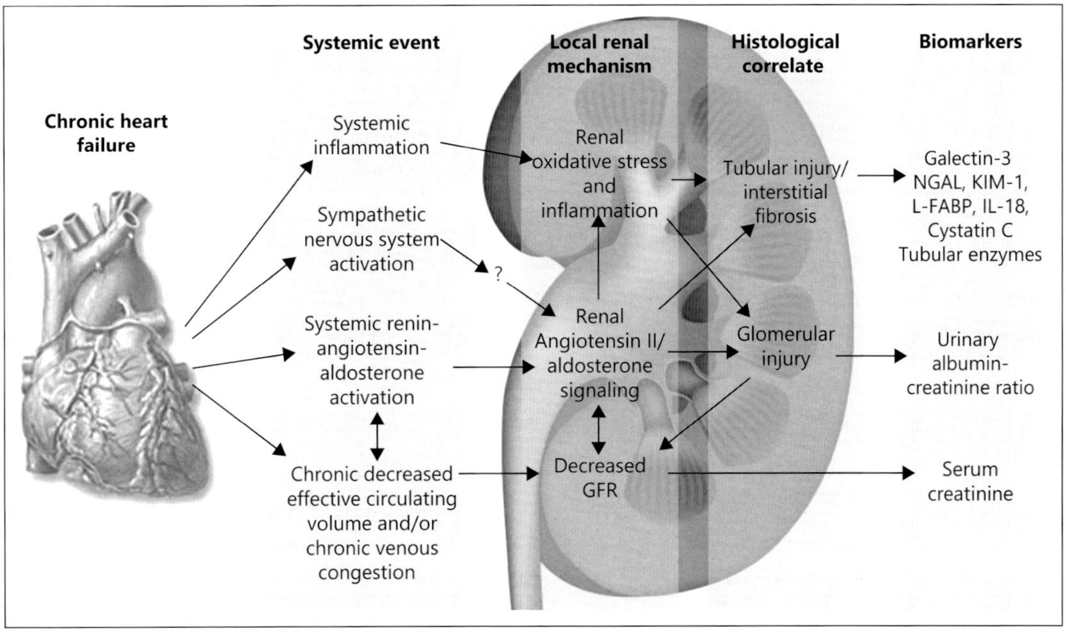

Fig. 4. Pathophysiology of CRS type 2.

late the transforming growth factor-β signal for increase cell cycle (cyclin D_1) enabling a marked production of procollagen I which is ultimately cross-linked into mature collagen [34]. Translational studies have demonstrated that a similar process of organ fibrosis can be turned on in the kidney resulting in progressive CKD in the setting of HF, warm renal ischemia-reperfusion injury, and renal transplantation degeneration [35, 36]. Thus, in CRS type 2, HF incites a sequence of processes involving the endothelium, glycocalyx, tubules, nephron units, and fibroblasts in the skeleton of the kidney that result in progressive loss of functioning nephrons and replacement of tissue with fibrosis (fig. 4) [37].

Acute Renocardiac Syndrome (Type 3): Acute Worsening of Kidney Function Leading to Heart Injury and/or Dysfunction

Little is known about the frequency of acute HF following AKI. It was inferred from the literature that subclinical HF or significant myocardial disease is an important determinant of the phenotypic expression of this syndrome [38]. Conversely, individuals with completely normal cardiac function rarely demonstrate HF in the setting of AKI. The mechanisms whereby AKI leads to cardiac dysfunction have been proposed to include two principles: direct effects of AKI on the heart, and effects of AKI on remote organ function with indirect effects on the heart (fig. 5) [39]. AKI has been shown to cause inflammation in experimental

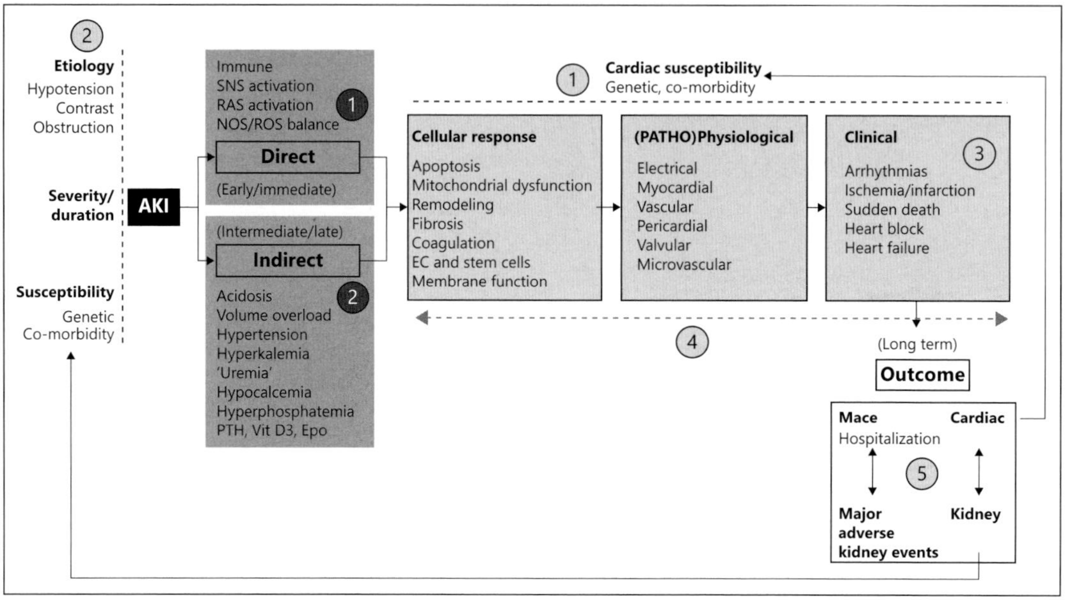

Fig. 5. Pathophysiology of CRS type 3.

renal ischemic models, which then induced cytokine expression, leukocyte infiltration into the heart, cell death by apoptosis, and impaired cardiac function [40]. Recent studies have shown that individuals with increased proximal tubular expression of kidney injury molecule-1, a marker of acute and CKD, have increased incidence of HF [41–43]. Combined with this finding are the well-known significant physiological derangements, including fluid and electrolyte imbalance and uremia that underpin remote organ failure and finally affect cardiac function, which in turn causes further kidney injury. Acute salt and water retention, volume overload, and the immediate effects of uremia on the myocardium are all postulated mechanisms in precipitating ADHF in CRS type 3 [44].

Chronic Renocardiac Syndrome (Type 4): CKD Leading to Heart Injury, Disease and/or Dysfunction

CKD is an accepted independent determinant for the progression of HF to hospitalization, pump failure death, and sudden death. In a recent meta-analysis, an exponential relation between the severity of renal dysfunction and the risk for all-cause mortality was described [45]. Overall mortality was driven by excess cardiovascular deaths, which constituted over 50% of the total mortality. CKD contributes to HF by causing or worsening the three fundamental mechanisms of cardiac failure: pressure overload, volume overload, and cardiomyopathy (fig. 6) [46]. Multiple studies have shown that CKD worsens hyperten-

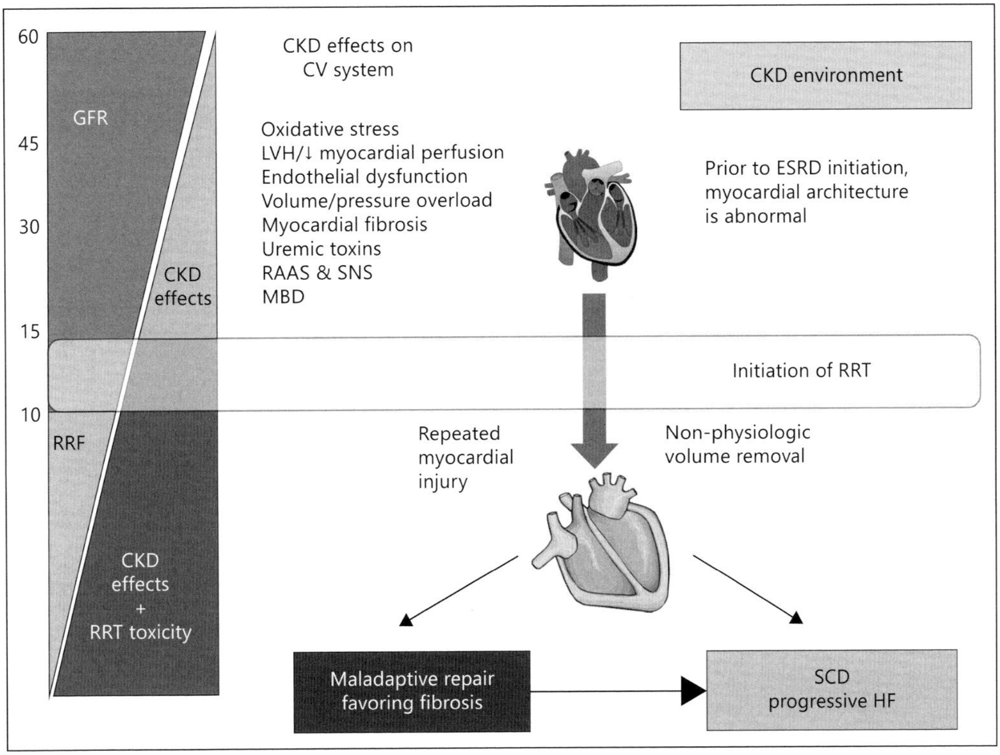

Fig. 6. Pathophysiology of CRS type 4. CKD = Chronic kidney disease; GFR = glomerular filtration rate; RRF = residual renal function; CV = cardiovascular; LVH = left ventricular hypertrophy; RAAS = renin angiotensin aldosterone system; SNS = sympathetic nerveous system, mineral and bone disease; ESRD = end-stage renal disease; RRT = renal replacement therapy; SCD = sudden cardiac death.

sion and impairs the success of antihypertensive agents [47]. Chronic salt and water retention increases volume loading on the cardiac chambers and results in elevated wall tension; this is reflected by chronic elevations of the natriuretic peptides produced by the myocardium [48]. In chronic uremia, elevation of multiple factors (marinobufagenin, indoxyl sulfate, *p*-cresol) have been shown to impair myocardial contractility, relaxation, and change the cellular and tissue characteristics of the myocardium [49–51]. As a result, uremia induces both systolic and diastolic dysfunction [52]. As renal filtration function declines, a reduction in the fractional excretion of phosphorus triggers osteocytes to produce fibroblast growth factor-23 (FGF-23), which stimulates FGF-23-Klotho receptor complexes in the kidneys facilitating greater degrees of fractional phosphate excretion while at the same time, having 'off-target' stimulating effects fibroblasts and myofibroblasts in the heart [53]. Both FGF-23 and a related protein, osteoprogerin, have been linked to HF in CKD patients [54]. Histopathologic changes in the myocardium in type 4 CRS include left ventricular hypertrophy, disassembly of gap junctions, increased expression of collagen-1, fibronectin, vimentin resulting in cardiac fibrosis, decreased capillary density, and increased tissue calcification [42, 55]. These changes constitute a 'CKD cardiomyopathy' and have been associated with both systolic and diastolic dysfunction, chronic edema, and increased rates of HF hospitalization and death (fig. 6). It is possible that given the large degree of left ventricular hypertrophy and reduction in capillary blood supply, that microvascular ischemia is a contributor to both symptoms and the natural history of CRS type 4. Chest pain in CKD patients has been associated with not only higher rates of acute coronary syndromes, but also the development of HF and hospitalization [56]. End-stage renal disease is the most exaggerated form of CRS type 4 and has been shown to relay added insults to the myocardium during chronic hemodialysis including myocardial stunning, tissue swelling, and marked increases in cell signaling and enzymatic activation of matrix metalloproteinase-2 [57]. The result of these changes include recruitment and activation of T cells and cardiac macrophages, which direct the cycle of cardiac myocyte apoptosis and replacement fibrosis as described above [58]. Cardiac fibrosis can be viewed as a final common substrate for the two major mechanisms of death in end-stage renal disease, pump failure and sudden arrhythmic death.

Secondary Cardiorenal Syndromes (Type 5): Systemic Conditions Leading to Simultaneous Injury and/or Dysfunction of Heart and Kidney

Although this subtype does not have a primary and secondary organ dysfunction, situations do arise where both organs simultaneously are targeted by systemic illnesses, the prototype being acute sepsis [59]. Sepsis is common and its incidence is increasing, with a mortality estimated between 20 and 60% [60]. Approximately 11–64% of septic patients develop AKI that is associated with a higher morbidity and mortality [61]. Abnormalities in cardiac function are also common in sepsis [62, 63]. Observational data have found approximately 30–80% of septic patients have elevated cardiac-specific troponin, that often correlate with reduced left ventricular function. The predominant mechanism is believed to be microcirculatory injury the results in elevations of cardiac troponin, natriuretic peptides, and evidence of left ventricular dysfunction along with various degrees of other organ failure (fig. 7) [64, 65]. At the same time, the cardiac features of AKI develop including azotemia and a reduction in urine output. Toxicity from markedly elevated neurohormones, particularly catecholamines, has been suggested in cases of overwhelming injury to the heart and kidneys, particularly in those with susceptible adrenergic receptor polymorphisms [66,

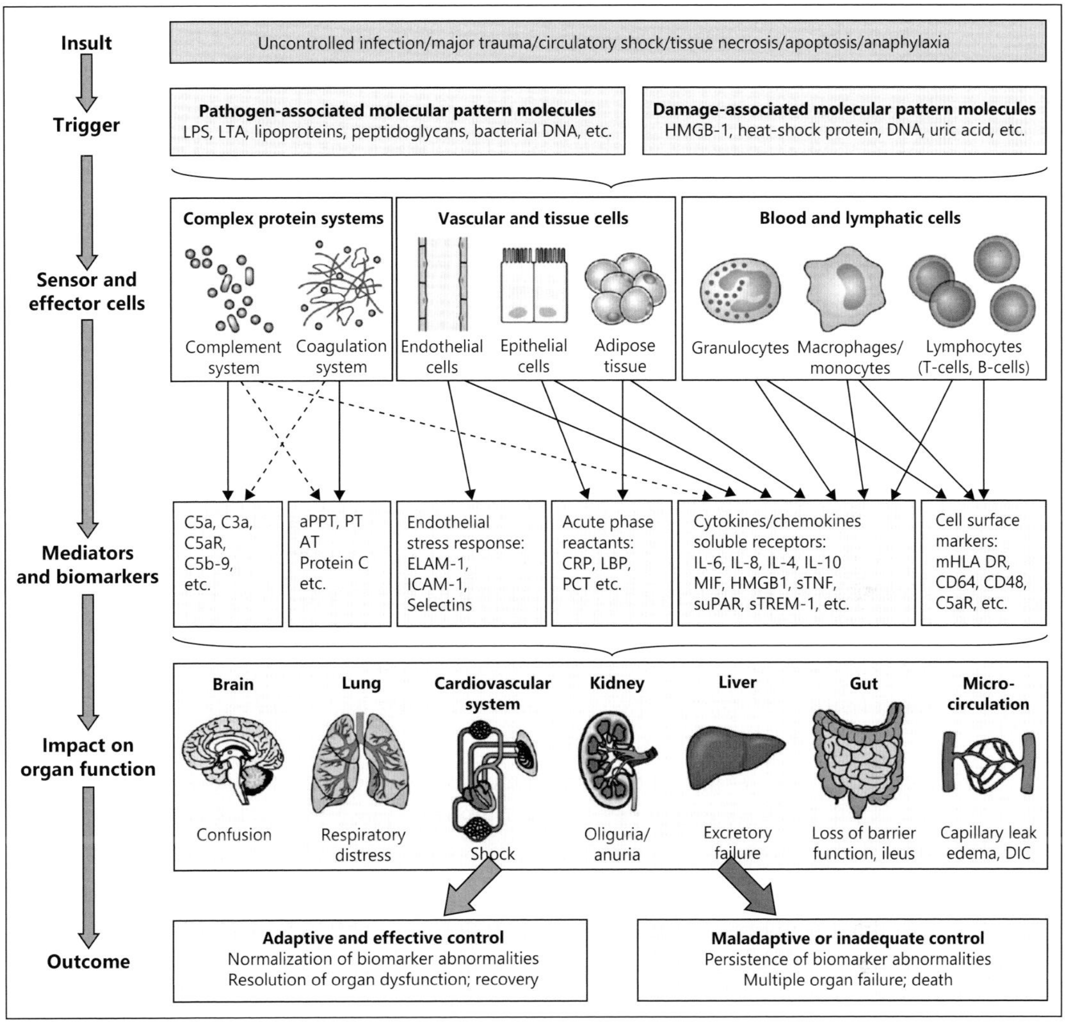

Fig. 7. Pathophysiology of CRS type 5. aPPT = Activated partial thromboplastin time; PT = prothrombin time; AT = activated clotting time; ELAM-1 = endothelial leukocyte adhesion molecule-1; ICAM-1 = intercellular adhesion molecule-1; CRP = C-reactive protein; LBP = lipopolysaccharide-binding protein; PCT = procalcitonin; IL = interleukin; MIF = macrophage migration inhibitory factor; HMGB1 = high mobility group box 1; sTNF = soluble tumor necrosis factor receptor; suPAR = soluble urokinase-type plasminogen activator receptor; sSTREM-1 = soluble triggering receptor expressed on myeloid cells; DIC = disseminated intravascular coagulation.

67]. When sepsis is complicated by shock, it is possible that ischemia/reperfusion injury occurs in both organs resulting in even further endothelial injury and impairment of microcirculatory perfusion [68, 69]. At the cellular and tissue level, abnormal cell signaling, cell cycle arrest, loss of control over oxidative reactions due to the liberation of catalytic iron from structural proteins (myoglobin) and mitochondria, acute cell death, apoptosis, and tissue necrosis can all be in the final common pathway in CRS type 5 [70]. Recently, myocardial cell Toll-like receptors (TL), particularly TL-3 on endosomes, has been implicated in the translation of cell signals in sepsis to myocyte necrosis [71]. In these cases, many other organs are often involved resulting in high in-hospital mortality [72]. In addition, if CRS type 5 survives, CKD and possibility chronic HF could be long-term outcomes of this viscous cycle [73].

Conclusions

Through the ADQI consensus on CRS, an integrative summary of leading pathophysiological considerations will lead to new thinking on novel diagnostic and therapeutic targets. There is a common need to understand acute and chronic hemodynamic mechanisms that regulate ventricular function, coronary flow, myocardial perfusion, systemic perfusion, and renal perfusion at multiple levels in the kidney. In addition, the impact of acute and chronic uremia, retention of nitrogenous and other molecules, and their impact on both the cardiac and renal systems needs more fundamental research. As renal filtration function improves, both cardiac and renal parameters indicate that retained toxins must play a role in combined pathogenesis. Abnormal regulation of neurohormonal systems appears to be a fundamental component of CRS pathophysiology. The degree of maladaptation and the relative response of counterregulatory mechanisms both appear to be fundamental in unlocking key diagnostic and therapeutic targets. Finally, immune cell signaling and enzymatic activation appear to be operative in many CRS, particularly those associated with acute and or chronic inflammation such as sepsis. Because both the heart and the kidneys have ongoing systemic demands, shifting of workloads from diseased or non-functional units to more functional ones appears to place an even greater burden of susceptibility to these remaining cardiomyocytes and nephrons as they work to maintain hemodynamic, metabolic, hormonal, nutritional, and other vital functions. As part of the final common pathway, tissue fibrosis in the heart and kidneys is apparent on histopathologic examination. Whether this is benign replacement of lost functional cells or directly pathogenic remains unknown. There are already several key diagnostic and therapeutic targets in clinical use and in development in this area. In summary, the pathophysiology of CRS is a complex array of mechanisms that will require considerable investigation from the research community to understand the major determinants and best pathways for diagnosis and management in the future.

Acknowledgements

The 11th International ADQI Consensus Conference was funded in part by generous support in form of unrestricted educational grants from Alere, Inc., Astute Medical, Inc., Gambro, and Fresenius Medical Care, Inc. The sponsors had no input on the content of this article.

References

1 Ledoux P: Cardiorenal syndrome (in French). Avenir Med 1951;48:149–153.

2 Blair JE, Manuchehry A, Chana A, Rossi J, Schrier RW, Burnett JC, Gheorghiade M: Prognostic markers in heart failure-congestion, neurohormones, and the cardiorenal syndrome. Acute Card Care 2007;9:207–213.

3 Sarraf M, Masoumi A, Schrier RW: Cardiorenal syndrome in acute decompensated heart failure. Clin J Am Soc Nephrol 2009;4:2013–2026.

4 Sarraf M, Schrier RW: Cardiorenal syndrome in acute heart failure syndromes. Int J Nephrol 2011;2011:293938.

5 Brammah A, Robertson S, Tait G, Isles C: Bilateral renovascular disease causing cardiorenal failure. BMJ 2003;326:489–491.

6 Bongartz LG, Cramer MJ, Doevendans PA, Joles JA, Braam B: The severe cardiorenal syndrome: 'Guyton revisited'. Eur Heart J 2005;26:11–17.

7 Ronco C, Haapio M, House AA, Anavekar N, Bellomo R: Cardiorenal syndrome. J Am Coll Cardiol 2008;52:1527–1539.

8 Ronco C, McCullough PA, Anker SD, Anand I, Aspromonte N, Bagshaw SM, Bellomo R, Berl T, Bobek I, Cruz DN, Daliento L, Davenport A, Haapio M, Hillege H, House A, Katz NM, Maisel A, Mankad S, Zanco P, Mebazaa A, Palazzuoli A, Ronco F, Shaw A, Sheinfeld G, Soni S, Vescovo G, Zamperetti N, Ponikowski P, Acute Dialysis Quality Initiative (ADQI) Consensus Group: Cardiorenal syndromes: an executive summary from the consensus conference of the Acute Dialysis Quality Initiative (ADQI). Contrib Nephrol. Basel, Karger, 2010, vol 165, pp 54–67.

9 Kellum JA, Bellomo R, Ronco C: Acute Dialysis Quality Initiative (ADQI): methodology. Int J Artif Organs 2008;31:90–93.

10 Bongartz LG, Braam B, Gaillard CA, Cramer MJ, Goldschmeding R, Verhaar MC, Doevendans PA, Joles JA: Target organ crosstalk in cardiorenal syndrome: animal models. Am J Physiol Renal Physiol 2012; 303:F1253–F1263.

11 Forman DE, Butler J, Wang Y, Abraham WT, O'Connor CM, Gottlieb SS, Loh E, Massie BM, Rich MW, Stevenson LW, Young JB, Krumholz HM: Incidence, predictors at admission, and impact of worsening renal function among patients hospitalized with heart failure. J Am Coll Cardiol 2004;43:61–67.

12 Gottlieb SS, Abraham W, Butler J, Forman DE, Loh E, Massie BM, O'Connor CM, Rich MW, Stevenson LW, Young J, Krumholz HM: The prognostic importance of different definitions of worsening renal function in congestive heart failure. J Card Fail 2002;8: 136–141.

13 Nohria A, Tsang SW, Fang JC, Lewis EF, Jarcho JA, Mudge GH, Stevenson LW: Clinical assessment identifies hemodynamic profiles that predict outcomes in patients admitted with heart failure. J Am Coll Cardiol 2003;41:1797–1804.

14 Haase M, Mueller C, Damman K, Murray P: Cardiorenal Syndrome Type 1 in Acute Decompensated Heart Failure: Workgroup Statements from the Eleventh Consensus Conference of the Acute Dialysis Quality Initiative (ADQI), 2013.

15 Ross EA: Congestive renal failure: the pathophysiology and treatment of renal venous hypertension. J Card Fail 2012;18: 930–938.

16 Tarantini L, Cioffi G, Gonzini L, Oliva F, Lucci D, Di Tano G, Maggioni AP, Tavazzi L, Italian Acute Heart Failure Survey: Evolution of renal function during and after an episode of cardiac decompensation: results from the Italian survey on acute heart failure. J Cardiovasc Med (Hagerstown) 2010;11:234–243.

56 McCullough PA, Nowak RM, Foreback C, Tokarski G, Tomlanovich MC, Khoury N, Weaver WD, Sandberg KR, McCord J: Emergency evaluation of chest pain in patients with advanced kidney disease. Arch Intern Med 2002;162:2464–2468.

57 Jefferies HJ, Virk B, Schiller B, Moran J, McIntyre CW: Frequent hemodialysis schedules are associated with reduced levels of dialysis-induced cardiac injury (myocardial stunning). Clin J Am Soc Nephrol 2011;6:1326–1332.

58 Han YL, Li YL, Jia LX, Cheng JZ, Qi YF, Zhang HJ, Du J: Reciprocal interaction between macrophages and T cells stimulates IFN-γ and MCP-1 production in Ang-II-induced cardiac inflammation and fibrosis. PLoS One 2012;7:e35506.

59 Soni SS, Ronco C, Pophale R, Bhansali AS, Nagarik AP, Barnela SR, Saboo SS, Raman A: Cardiorenal syndrome type 5: epidemiology, pathophysiology, and treatment. Semin Nephrol 2012;32:49–56.

60 Alberti C, Brun-Buisson C, Burchardi H, Martin C, Goodman S, Artigas A, Sicignano A, Palazzo M, Moreno R, Boulme R, Lepage E, Le Gall R: Epidemiology of sepsis and infection in ICU patients from an international multicentre cohort study. Intensive Care Med 2002;28:108–121.

61 Hoste EA, Lameire NH, Vanholder RC, Benoit DD, Decruyenaere JM, Colardyn FA: Acute renal failure in patients with sepsis in a surgical ICU: predictive factors, incidence, comorbidity, and outcome. J Am Soc Nephrol 2003;14:1022–1030.

62 Charpentier J, Luyt CE, Fulla Y, Vinsonneau C, Cariou A, Grabar S, Dhainaut JF, Mira JP, Chiche JD: Brain natriuretic peptide: a marker of myocardial dysfunction and prognosis during severe sepsis. Crit Care Med 2004;32:660–665.

63 Ver Elst KM, Spapen HD, Nguyen DN, Garbar C, Huyghens LP, Gorus FK: Cardiac troponins I and T are biological markers of left ventricular dysfunction in septic shock. Clin Chem 2000;46:650–657.

64 Mehta NJ, Khan IA, Gupta V, Jani K, Gowda RM, Smith PR: Cardiac troponin I predicts myocardial dysfunction and adverse outcome in septic shock. Int J Cardiol 2004;95:13–17.

65 Chelazzi C, Villa G, De Gaudio AR: Cardiorenal syndromes and sepsis. Int J Nephrol. 2011;2011:652967.

66 Mohaupt M, Kramer HJ: Acute ischemic renal failure: review of experimental studies on pathophysiology and potential protective interventions. Ren Fail 1989–1990;11:177–185.

67 Vriz O, Minisini R, Citro R, Guerra V, Zito C, De Luca G, Pavan D, Pirisi M, Limongelli G, Bossone E: Analysis of β_1- and β_2-adrenergic receptors polymorphism in patients with apical ballooning cardiomyopathy. Acta Cardiol 2011;66:787–790.

68 Celes MR, Prado CM, Rossi MA: Sepsis: going to the heart of the matter. Pathobiology 2013;80:70–86.

69 Mehta RL, Rabb H, Shaw A, Singbartl K: Cardiorenal Syndrome Type 5: Clinical Presentation, Pathophysiology and Management Strategies. Summary of Recommendations from the Acute Dialysis Quality Initiative Consensus Conference, 2013.

70 Rudiger A, Dyson A, Felsmann K, Carré JE, Taylor V, Hughes S, Clatworthy I, Protti A, Pellerin D, Lemm J, Claus RA, Bauer M, Singer M: Early functional and transcriptomic changes in the myocardium predict outcome in a long-term rat model of sepsis. Clin Sci (Lond) 2013;124:391–401.

71 Gao M, Ha T, Zhang X, Liu L, Wang X, Kelley J, Singh K, Kao R, Gao X, Williams D, Li C: Toll-like receptor 3 plays a central role in cardiac dysfunction during polymicrobial sepsis. Crit Care Med 2012;40:2390–2399.

72 Ferreira AM, Sakr Y: Organ dysfunction: general approach, epidemiology, and organ failure scores. Semin Respir Crit Care Med 2011;32:543–551.

73 Chelazzi C, Villa G, De Gaudio AR: Cardiorenal syndromes and sepsis. Int J Nephrol 2011;2011:652967.

Peter A. McCullough, MD, MPH
Providence Park Heart Institute, Providence Park Hospital
47601 Grand River, B125
Novi, MI 48374 (USA)
E-Mail peteramccullough@gmail.com

McCullough PA, Kellum JA, Mehta RL, Murray PT, Ronco C (eds): ADQI Consensus on AKI Biomarkers and Cardiorenal Syndromes. Contrib Nephrol. Basel, Karger, 2013, vol 182, pp 99–116
DOI: 10.1159/000349969

Pathogenesis of Cardiorenal Syndrome Type 1 in Acute Decompensated Heart Failure: Workgroup Statements from the Eleventh Consensus Conference of the Acute Dialysis Quality Initiative (ADQI)

Michael Haase[a] · Christian Müller[b] · Kevin Damman[c] · Patrick T. Murray[d] · John A. Kellum[e] · Claudio Ronco[f] · Peter A. McCullough[g] · for the Acute Dialysis Quality Initiative (ADQI) Consensus Group

[a]Department of Nephrology and Hypertension, Diabetes and Endocrinology, Otto von Guericke University Magdeburg, Magdeburg, Germany; [b]Department of Cardiology, University Hospital, Basel, Switzerland; [c]Department of Cardiology, University Medical Center Groningen, University of Groningen, Groningen, The Netherlands; [d]Nephrology and Clinical Pharmacology, University College Dublin School of Medicine and Medical Science, Dublin, Ireland; [e]Clinical Research, Investigation, and Systems Modeling of Acute Illness Center, Department of Critical Care Medicine, University of Pittsburgh, Pittsburgh, Pa., USA; [f]Department of Nephrology, International Renal Research Institute (IRRIV), San Bortolo Hospital, Vicenza, Italy; [g]St. John Providence Health System, Warren, Mich., Providence Hospitals and Medical Centers, Southfield and Novi, Mich., St. John Macomb Oakland Center, Madison Heights, Mich., St. John Hospital and Medical Center, Detroit, Mich., USA

Abstract

Pathophysiological mechanisms of cardiorenal syndromes (CRS) types 1–5 are still sparsely characterized. In an attempt to address this issue, a consensus conference on CRS was held in Venice, Italy, in November 2012 under the auspices of the Acute Dialysis Quality Initiative (ADQI). Working group 1 discussed monodirectional mechanisms of CRS type 1 which is characterized by a rapid worsening of cardiac function leading to acute kidney injury (AKI). Pre-conference we performed a systematic search and review of the available literature using a modified Delphi analysis. Hereby identified and in this review discussed questions were: (i) What are the predominant pathophysiologic mechanisms of CRS type 1 in acute decompensated heart failure? (ii) Could biomarker profiling identify pathomechanisms or hemodynamic phenotype of patients with CRS type 1? Could predictive biomarkers improve renal safety of therapy in CRS type 1? (iii) How do the timing, severity

ADQI 11 Workgroup members are listed in Appendix 2.

and duration relate to the mechanisms and outcomes of CRS type 1? In summary, after discussion and appraisal of the best available evidence, working group 1 makes consensus recommendations for future research on pathologic mechanisms of CRS type 1 and recommendations for clinical practice where treatment is in either proof or disproof of a mechanism.

Bidirectional acute and chronic disorders of heart and kidney are today classified as cardiorenal syndromes (CRS) types 1–5, however pathophysiological mechanisms are sparsely characterized with regard to correct identification and treatment at clinical presentation. In an attempt to address this issue, a consensus conference on CRS was held in Venice, Italy, in November 2012 under the auspices of the Acute Dialysis Quality Initiative (ADQI).

Working group 1 addressed these issues for CRS type 1 which is characterized by a rapid worsening of cardiac function leading to acute kidney injury (AKI) [1, 2]. CRS type 1 most frequently appears in the setting of acute decompensated heart failure (ADHF) [3] and follows ischemic (cardiac surgery, myocardial infarction) or non-ischemic (valve dysfunction, aortic dissection, pulmonary embolism, etc.) cardiac events. Up to 40% of patients hospitalized for ADHF develop AKI [4]. These patients require a more complex management and experience more complicated hospital courses and higher mortality. Preliminary data point towards the importance of timing in the development of AKI with regard to severity of the renal stressor versus iatrogenic effects.

Classical mechanisms of CRS type 1 include low cardiac output (CO) and neurohormonal activation and release of vasoactive substances resulting in low renal perfusion and possible renal ischemia with AKI. In addition, high central venous pressure (CVP), increased intra-abdominal pressure leading to venous congestion, activation of the sympathetic nervous system and the renin-angiotensin-aldosterone system (RAAS) and release of other vasoactive substances such as endothelin, anemia and a marked alteration of immune and somatic cell signaling have all been implicated as important contributors of kidney injury (fig. 1).

As far as understanding CRS type 1 pathophysiology is concerned, treatment approaches including agents targeting vasoactive and neurohormonal pathways and strategies to correct renal congestion through negative fluid balance are briefly discussed.

Methods

Pre-conference we performed a systematic search and review of the available literature, as described in the ADQI 11 conference summary statements elsewhere in this issue. Specifically, we focused on the identification of pathomechanisms of the acutely dys-

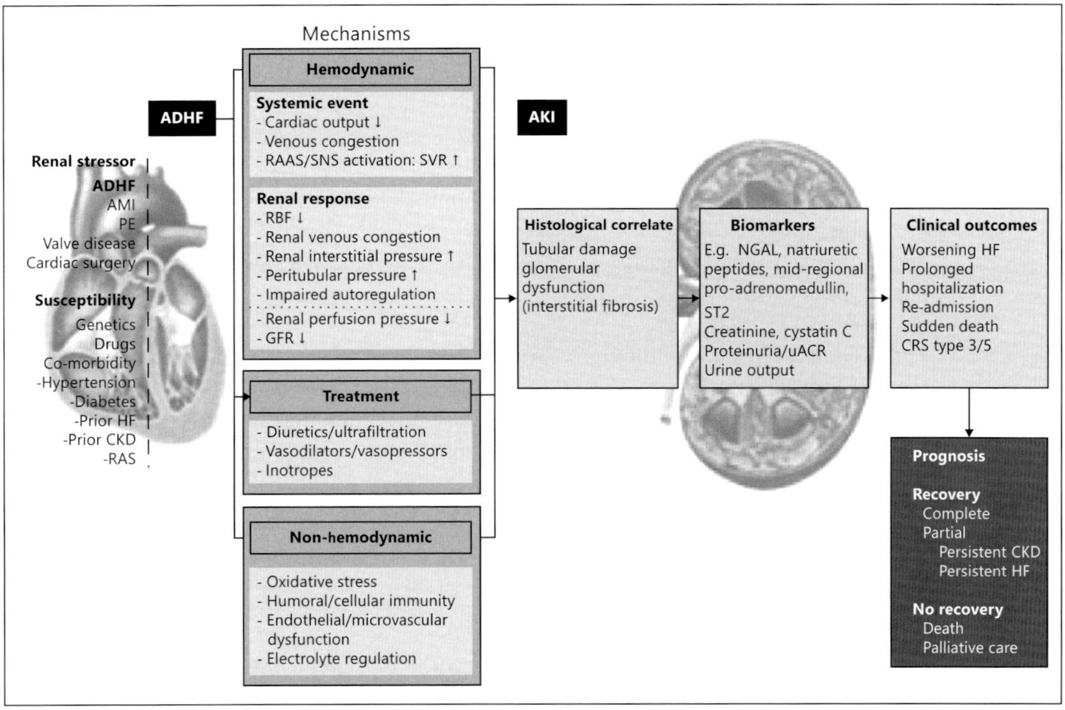

Fig. 1. Overview on mechanisms, histological correlates biomarkers and outcomes in CRS type 1 in the setting of acute decompensated heart failure. Reproduced with permission from ADQI [53]. ADHF = Acute decompensated heart failure; AKI = acute kidney injury (the term AKI also covers the term 'WRF', 'worsening renal function which is usually defined as doubling in serum creatinine); AMI = acute myocardial infarction; PE = pulmonary embolism; SVR = systemic vascular resistance; HF = heart failure; CKD = chronic kidney disease; RAS = renal artery stenosis; GFR = glomerular filtration rate; NGAL = neutrophil gelatinase-associated lipocalin; IL-18 = interleukin-18; KIM-1 = kidney injury molecule 1; L-FABP = liver-type fatty acid binding protein; NAG = N-acetylglucosamine.

functioning heart damaging the kidney and their potential to be correctly identified in daily practice. Treatment effects in CRS type 1 were only considered either as proof or disproof of a mechanism. Experiments and studies were identified via Medline and Web of Science using the terms 'cardiorenal', '(acutely/acute decompensated) heart failure', and 'acute kidney injury (AKI)' or '(worsening) renal (function)' combined with 'animal (experiment/model)', '(patho)mechanism', or 'pathophysiology', 'pathway' or 'RAAS', 'fluid retention', 'sympathetic nervous system', 'biomarker' or 'immunity'. In view of the volume of the retrieved literature, only representative publications are cited in this review. Furthermore, we opted to limit this summary to clinical scenarios in which there was ample data. We did not include the specific clinical situations of cardiac surgery procedures with the use of cardiopulmonary bypass or contrast media application as we rank both scenarios as antecedents to CRS type 1.

This report is the result of a modified Delphi analysis [5] which is a structured and standardized process for collecting, summarizing and disseminating knowledge from a group of experts focused on a specific problem or task. A detailed description of the ADQI methodology is available at www.adqi.net.

Based on the literature identified prior to the conference, the following key questions were identified:

Question 1: What are the predominant pathophysiologic mechanisms of CRS type 1 in ADHF?

Question 2: Could biomarker profiling identify pathomechanisms or hemodynamic phenotype of patients with CRS type 1? Could predictive biomarkers improve renal safety of therapy in CRS type 1?

Question 3: How do the timing, severity and duration relate to the mechanisms and outcomes of CRS type 1?

At the conference, after a systematic literature review and the appraisal of the best available evidence, group 1 discussed monodirectional mechanisms of CRS type 1 generating from the heart affecting the kidney.

Results

Ad Question 1: What are the predominant pathophysiologic mechanisms of CRS type 1 in ADHF?

Hemodynamic Mechanisms

Observations from experiments and the clinical setting suggest that hemodynamic mechanisms seem to play a major if not predominant role in CRS type 1 in the setting of ADHF. Various animal studies suggest an initial hemodynamic event as a CRS type 1 trigger leading to decreased renal arterial flow, renal oxygen consumption and glomerular filtration rate (GFR) and increased renal vascular resistance whereas hemodynamic counteractive measures such as renal-selective blood perfusion restored all of these hemodynamic parameters and renal function back to normal [6].

In the general setting of ADHF, different hemodynamic profiles have been proposed based on the clinical phenotype of individual patients [7]. This approach consists of categorization of patients dependent of their systemic hemodynamics, including adequacy of perfusion (decreased CO and decreased effective circulation fluid volume (ECFV)) and extent of (pulmonary) congestion (increase in CVP or wedge pressure). As such, these can be combined into four distinct profiles deemed wet or dry and warm or cold (fig. 2).

It is within each of these profiles of ADHF where CRS type 1 may develop. Since hemodynamics, patient characteristics, treatment, and even outcome

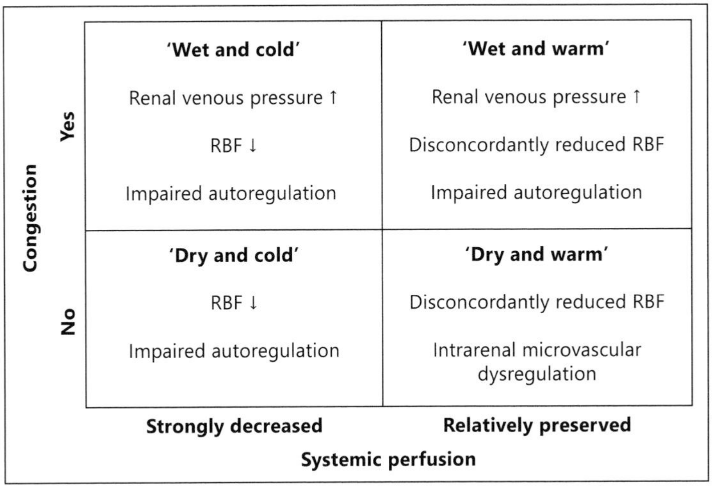

Fig. 2. Systemic hemodynamic profiles of ADHF and renal hemodynamic consequences in CRS type 1. Adapted from Stevenson and Perloff [7]. This figure combines profiles of ADHF with systemic hemodynamics and possible hemodynamic renal mechanisms that can cause CRS type 1 in each of the profiles. In the 'cold' profiles, AKI may develop as a consequence of reduced RBF when autoregulation is unable to preserve GFR. In the 'dry and cold' profile, this may include anemia or stringent diuresis/ultrafiltration with associated blood pressure reductions. In the 'wet' profiles, increased central venous pressure leads to increased renal venous pressure, that reduces renal perfusion pressure, increases renal interstitial pressure, opposing filtration by collapse of tubules. In the 'warm' profiles, although systemic perfusion is relatively preserved, RBF is disconcordantly reduced, attributable to impaired or attenuated autoregulation (e.g. RAAS inhibition, excessive sympathetic nervous system activation), impaired tubuloglomerular feedback mechanisms (e.g. NSAIDs), renal artery stenosis or predisposing conditions with loss of nephrons. Note that in each of the profiles, impaired (or attenuated) autoregulation may play a key role, resulting in decreased renal perfusion pressure. Considering the fact that most ADHF patients have high blood pressure, if CRS type 1 develops as a consequence of decreased RBF (and/or decreased renal perfusion pressure), this must be at least partly attributable to autoregulatory dysfunction. ADHF = Acute decompensated heart failure; CRS = cardiorenal syndrome; AKI = acute kidney injury; RBF = renal blood flow; GFR = glomerular filtration rate; RAAS = renin-angiotensin-aldosterone system; NSAID = non-steroidal inflammatory drugs.

are different in each of the four profiles, it is likely that also the pathophysiology of CRS type 1 may differ dependent of the hemodynamic profile of ADHF [8]. In addition, the occurrence of AKI in this setting may not only be restricted to within a certain profile, but may also be related to the dynamic process of shifting between profiles during either treatment or deterioration of ADHF.

Hemodynamic Profiles of CRS Type 1 in the Setting of ADHF

Decreased Systemic Perfusion ('Cold' Profiles)

(A) Reduced Renal Blood Flow

In the profiles that both have a cold phenotype (fig. 2), the predominant alteration in systemic hemodynamics is a strong reduction in CO and ECFV, and this may be accompanied by marked increase in CVP in the wet profile. When CRS type 1 develops in these 'cold' patients, it is more than likely associated with a reduction in renal blood flow (RBF). In an acute setting such as ADHF it is however unclear how CO relates to RBF or even intrarenal blood flow distribution in ADHF. It may be hypothesized however that with a strong decrease in ECFV as is to be expected in ADHF, neurohormonal activation including renin-angiotensin system and systemic nervous system activation result in afferent (and relatively lower efferent) vasoconstriction, leading to a decrease in RBF and renal perfusion pressure. In these 'cold' profiles, low CO and ECFV may also be associated with low systemic blood pressure. Consequently, this reduces renal perfusion pressure, even in the case of relatively normal renal venous pressure, because autoregulation may be unable to compensate for the low blood pressure in the presence of low ECFV. The finding of severely decreased RBF and GFR in the setting of reduced CO would therefore imply that autoregulation is either (i) unable to compensate, (ii) attenuated or (iii) overruled by different mechanisms, or (iv) failing because of an unknown cause. In any case, data on these mechanisms in the setting of ADHF and CRS type 1 are lacking.

(B) Increased Central Venous Pressure

For those patients who are also 'wet' and therefore have increased systemic/pulmonary congestion, different mechanisms may cause CRS type 1. High CVP is directly transmitted to the renal vein and directly influences renal perfusion pressure. Different reports have highlighted that higher CVP (either caused by intravascular hypertension or by increased intra-abdominal pressure) is associated with decreasing GFR [9–12]. Next to a direct effect on renal perfusion pressure, high renal venous pressure results in increased interstitial intrarenal pressure because the kidney has a tight capsule. This increased pressure causes collapsing of tubules, and directly opposes filtration, resulting in decreased GFR [13]. How autoregulation responds to increased renal venous pressure is unknown, although higher levels of intrarenal angiotensin II and activation of the sympathetic nervous system have been proposed, which could indirectly influence arteriolar tone [14, 15].

(Relatively) Preserved Perfusion ('Warm' Profiles)
It is essential to highlight that even in the 'warm' profiles, CO and ECFV are still reduced compared to normal values, but are relatively high compared to those in the cold profiles. It might therefore be difficult to imagine how CRS type 1 may develop as a consequence of decreased RBF. In these specific patient groups, there is no data on RBF. If CO is relatively preserved, but RBF is decreased to such an extent that is associated with a low GFR, this implies there is a disproportionate decrease in RBF compared to CO. These circumstances may develop when uni- or bilateral renal artery stenosis is present, which has been estimated to be present up to 40% in patients with chronic heart failure and chronic kidney disease [16]. Even a small reduction in CO would result in larger reductions in RBF. Other factors that may be particularly important as pathophysiologic mechanisms of CRS type 1 in these profiles are chronic use of RAAS inhibitors that limit the autoregulatory response to reductions in ECFV and associated blood pressures. In a similar fashion, concomitant non-steroidal inflammatory drug (NSAID) use may limit tubuloglomerular feedback (TGF) when in response to a decreased ECFV TGF activation causes afferent vasodilation, which may be blocked by NSAIDs. Also, a small decrease in CO may not only result in an overall reduction in RBF, but can also change intrarenal blood flow distributions, of which the importance and effect on GFR in ADHF is unknown. Finally, since patients in the warm profiles tend to present with high blood pressures, the co- and preexisting hypertension in the chronic disease state can be associated with a reduction in functional nephrons.

The warm/wet profile with increased CVP is the most frequently occurring profile in acute and advanced HF [8], and therefore the possibility that CRS type 1 occurs in especially this setting is higher beforehand compared to any other. Mechanisms of increased CVP in this profile leading to AKI are not different compared to those in the cold profiles. However, the relative impact of increased CVP on renal perfusion pressure will be much more limited, since arterial blood pressures in this profile are usually high. A small increase in CVP will therefore have less impact on renal perfusion pressure. Whether high CVP relates to lower GFR and more frequent AKI in patients with high or low CO is matter of debate. Mullens et al. [9] showed that AKI especially occurred in those with high CI and high CVP. Damman et al. [10] showed in a variety of patients with cardiovascular disease that higher CVP was associated with lower eGFR even in patients with higher CI. On the other hand, increased CVP was not associated with AKI in the ESCAPE trial [11], and lower CVP predisposed to AKI in one study [12]. Possibly, it is especially this category of patients who will be treated rigorously with high dosages of diuretics (or ultrafiltration). As a consequence, if this treatment is paralleled by decreases in (arterial) blood pressures, it is more frequently associated with AKI and poor outcome.

Consensus Recommendations for Future Research

- Appropriate experimental models and clinical scenarios for CRS type 1 need to be identified; combined heart/kidney transplantation with intraoperative biopsy would need to be considered.
- Prospective and systematic studies on the incidence, mechanisms and associated outcome of CRS type 1 in different hemodynamic profiles of ADHF are needed.
- Therapy aimed at restoring systemic perfusion and reducing congestion without further reducing renal perfusion pressure should be a therapeutic goal.
- Relative importance of (non-invasive) RBF in ADHF and the development of CRS type 1 need to be determined.
- Relative importance of central and renal venous pressure in ADHF and the development of CRS type 1 need to be determined.

Consensus Recommendations for Clinical Practice

- Identification of hemodynamic profiles in patients with ADHF and CRS type 1 may help in guiding therapy decisions.

Non-Hemodynamic Mechanisms

Beyond hemodynamic pathways, other mechanisms have been described to be involved in pathogenesis of CRS type 1 including activation of sympathetic nervous system and RAAS, inflammation and impaired balance of ROS versus NO production.

Systemic hemodynamic events occurring during ADHF contribute to decreased RBF, impaired autoregulation and renal congestion (fig. 2) with the latter resulting from the product of right heart function, blood volume, and venous capacity, which is largely regulated by neurohormonal systems. Specific pathobiological (back)coupling mechanisms and modifiers are activated with positive or negative feedback loops to continuously monitor and adapt to changing extracellular fluid volume and blood pressure, depending on the duration and the intensity of the insult (fig. 3).

In the setting of significant acute heart dysfunction, falling CO and decreasing ECFV of the renal arterial tree results in activation of both SNS and RAAS mediated for example through the baroreceptor reflex. Also, increased central

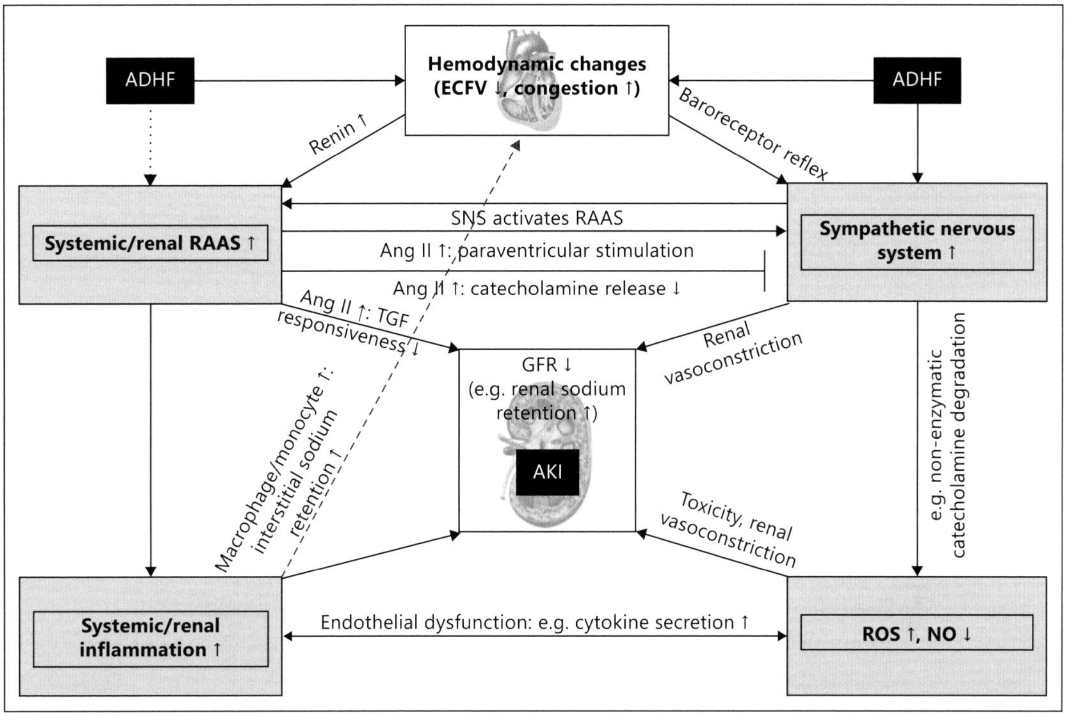

Fig. 3. Non-hemodynamic network of pathophysiological interactions in CRS type 1. Note the emerging potential role of macrophages/monocytes as mediator of sodium and fluid retention. Reproduced with permission from ADQI [53].

stimulation of the SNS in the paraventricular nucleus – presumably by a combination of factors including increased angiotensin II, decreased NO and increased afferent renal sympathetic nerve activity – and extracellular volume expansion/control contribute to alterations in GFR and RBF in CRS type 1 [13].

Recent data suggest a potential role of mononuclear phagocyte system in the regulation of salt-dependent volume and blood pressure by a vascular endothelial growth factor-dependent mechanism [17]. Specifically, macrophages may be protective against the development of arterial hypertension and volume overload. By implication, a dysregulation of macrophages may initiate the sequence of ADHF through volume overload and congestion even without a primary hemodynamic cardiovascular event making macrophage modulation a potentially interesting target in CRS type 1.

In line with cellular immunity potentially being critically involved in the pathogenesis of human CRS type 1, Virzi et al. [18] report that a defective regulation of monocyte apoptosis and activation of inflammatory pathways occurs more frequently in patients with ADHF compared to healthy controls. Plasma-induced apoptosis was 7-fold higher in CRS type 1 patients, the activity of caspase-8 was significantly higher in monocytes incubated with plasma from affected patients and the serum concentration of two pro-inflammatory cytokines was elevated 4- to 10-fold. Therefore, immune cell apoptosis in CRS type 1 and with it premise of an immune-mediated process in the pathophysiology of CRS type 1 should be further explored.

Consensus Recommendations for Future Research

- Useful experimental and clinical models for CRS type 1 delineating bidirectional heart-kidney interactions need to be identified.
- Prospective and systematic studies with simultaneous measurement of hemodynamic and non-hemodynamic indices to determine the relative importance of pathomechanisms in initiation and maintenance of CRS type 1 in different hemodynamic profiles of ADHF are needed.
- The role of cellular immunity and resulting sodium and fluid retention in initiation and maintenance of CRS type 1 need to be further explored in clinical studies.

Consensus Recommendations for Clinical Practice

- At this stage there is no evidence to initiate specific anti-inflammatory or anti-neurohormonal treatment in CRS type 1.

Ad Question 2: Could biomarker profiling identify pathomechanisms or hemodynamic phenotype of patients with CRS type 1? Could predictive biomarkers improve renal safety of therapy in CRS type 1?

The challenge in management of CRS type 1 is relieving congestion and improving symptoms while restoring renal and systemic perfusion. Although pathophysiological understanding of ADHF and CRS type 1 increases, morbidity and mortality of the syndrome remained high over the last decades. Multiple reasons for failure of effective therapies have been implicated including the fact that (i) multiple causative factors may be involved and tackling of one pathway may be insufficient, (ii) single intervention may not cover the entire pathophysiological process, (iii) the time window between insult and development of AKI can be different in different patients, (iv) testing of drugs during their development may not reproduce their possible use in multimorbid patients, and (v) in the clinical setting AKI is often diagnosed too late, when the effects of the insult already become evident (loss of renal excretory function).

Therefore, earlier and more specific diagnosis of CRS type 1 but also allocation of predominant mechanism, site of renal injury and hemodynamic phenotype in an individual patient is needed. In this regard, biochemical cardiorenal markers, imaging techniques and monitoring of fluid and hemodynamic status will be briefly discussed.

Two important characteristics of biomarkers should be their defined role in the pathophysiology of the syndrome and their clinical actionability. Regarding the first postulated characteristic, neutrophil gelatinase-associated lipocalin (NGAL) has been implicated in the induction of cardiomyocyte apoptosis by increasing intracellular iron accumulation [19]. Also, administration of recombinant NGAL to mice induced an acute inflammatory response with compensatory changes in cardiac functional parameters reflecting its potential role as cardiorenal biomarker [19]. Such findings were in line with those reported for experimental and human myocarditis where NGAL was strongly induced in affected myocardiocytes, vascular wall cells, fibroblasts and neutrophils [20]. High NGAL levels were initially sustained during the inflammatory stage, then decreased with recovery [20]. Beyond a potential inflammatory role of NGAL in CRS, there is first evidence for its involvement in fluid status regulation given that mineral corticoid receptor activation induces NGAL promoter in the cardiovascular system and upregulation of NGAL expression in the heart and aorta and its plasma levels [21].

Biomarker evidence of neurohormonal activation includes the natriuretic peptides, mid-regional pro-adrenomedullin, and copeptin [22, 23]. Recent studies suggest individuals with greater levels of natriuretic peptides as baseline and evidence of pulmonary congestion, in addition to increased CVP, have the greatest risk for CRS type 1 [24]. Such hemodynamic impairment of the kidney may result in a loss of autoregulation and the onset of worsened salt and water retention, reduction in renal filtration, and oliguria.

Beyond hemodynamics and neurohormonal activation, abnormal patterns of cell signaling have been linked to ADHF and possibly CRS type 1. Soluble ST-2, a recently approved blood biomarker in the risk prediction of HF hospitalization and death [25, 26], appears to be complementary to the natriuretic peptides and reflects the degree of biomechanical strain and immune cell activation and cell signaling that occurs in progressive HF [27]. ST2 is the receptor for interleukin (IL)-33, a cytokine with angiohypertrophic and antifibrotic effects on the myocardium [28]. Soluble ST is a circulating inhibitor of the ST2 receptor, and counteracts such potentially regenerative effects of IL-33. This line of reasoning illustrates how systemic cell signaling could trigger events in the kidney in the form of maladaptive TGF which cause a transient reduction in renal filtration and a rise in serum creatinine [29, 30].

The use of biochemical markers for detection of loss of GFR and development of acute tubular damage has been suggested to overcome limitations of serum creatinine. In this regard, ADQI recommended a novel unified diagnostic concept with the biochemical detection of acute tubular damage complementing acute loss of renal excretory function [31].

Beside serum creatinine, serum cystatin C, a 25-kDa glomerularly filterable protein produced by all nucleated cells at a constant rate, has been used as biochemical surrogate for renal function. Serum cystatin C measured at admission in patients hospitalized for treatment of ADHF was more predictive of long-term all-cause mortality and readmission for ADHF than serum creatinine or serum BNP [32–34]. Given the prognostic value and its biochemical characteristics, serum cystatin C may also have value for earlier diagnosis of an acute loss of renal excretory function in CRS type 1, however data in this regard are lacking. Drawing pathophysiological inferences from the serum cystatin C test result about allocation to one out of four above-mentioned specific hemodynamic phenotypes does not appear to be possible.

Among other tubular damage markers such as kidney injury molecule 1 (KIM-1), liver-type fatty acid binding protein (L-FABP), IL-18 and others, NGAL appears to be that one with most comprehensive experimental and clinical data available in ADHF at this stage. NGAL levels were found to be largely determined by underlying impairment of renal rather than myocardial function [35]. In patients with ADHF, serum NGAL strongly correlated with renal function markers [36]. Higher serum NGAL levels at hospital admission have been associated with adverse cardiovascular outcomes or death [37]. Of note, at hospital discharge, plasma NGAL level was a stronger predictor of 30-day all-cause death and ADHF readmissions than BNP (adjusted HR for NGAL: 19.9 vs. adjusted HR for BNP: 2.3) [38]. Such comparison with BNP suggests that ongoing acute tubular damage in CRS type 1 is of importance for future development of CRS type 3.

There is preliminary evidence that elevated admission serum NGAL levels predict AKI in patients with ADHF whereas serum cystatin C and even more serum creatinine were of limited value [39–41]. One study compared the value of urine NGAL with that of plasma NGAL for predicting AKI in patients with ADHF and found that both markers were similar in this regard [42].

In one study, NGAL raised slightly during diuretic treatment of ADHF in patients who subsequently developed AKI. However, NGAL predictive ability for AKI in this setting was limited [43].

Further candidate biomarkers potentially related to pathophysiology of CRS type 1 and to be tested for suitability for the early diagnosis of AKI in patients with ADHF include osteopontin, N-acetyl-β-D-glucosaminidase, stromal cell-derived factor-1, endoglin, galectin-3 and exosomes [44]. However, at this stage, data are too limited to draw any conclusions for use of these candidate biomarkers.

A different approach from biochemical biomarkers in the differential diagnosis of CRS type 1 is the use of bioelectrical devices to measure partial or whole-body impedance and conclude from such value on the fluid status and with this allocation of a hemodynamic profile of a given patient as described above. In the absence of comprehensive data on the clinical usefulness of bioimpedance measurement in CRS type 1, several clinical studies in patients presenting with ADHF showed an association of decreased impedance values, i.e. increased body fluid volume, with adverse events including rehospitalization and death. Also, in patients presenting with dyspnea, bioimpedance measurement helped to distinguish cardiogenic dyspnea from non-cardiogenic dyspnea [45, 46]. Sequential bioimpedance measurements may contribute to achievement of adequate fluid balance status in patients with ADHF [47]. High sensitivity and specificity of reduced bioimpedance value were reported for identifying radiographic findings consistent with pulmonary edema [48]. Bioimpedance test result may be a potential criterion for discharge readiness [49].

Taken together, if acute tubular damage markers contribute to distinguishing a potentially adaptive (compensation for reduced ECFV) from a maladaptive reaction of the kidney in response to ADHF, their measurement in clinical practice might prove useful. Such approach would consider rapid reversibility of renal function loss without tubular damage. Bioimpedance measurement may guide early diagnosis of ADHF, differential diagnosis of dyspnea and therapy management on CRS type 1 but need further exploration.

Consensus Recommendations for Future Research

- Future studies should investigate the role of mineralcorticoid receptor antagonists in CRS type 1.
- Multicenter studies using a panel of acute tubular damage markers measured in urine and in plasma at admission and discharge and allocation to hemodynamic phenotype/profile are needed.
- The role of bioimpedance measurement for prevention and/or therapy of CRS type 1 needs to be investigated.

Consensus Recommendations for Clinical Practice

- Use of bioimpedance measurement may be considered in the management of patients with CRS type 1.

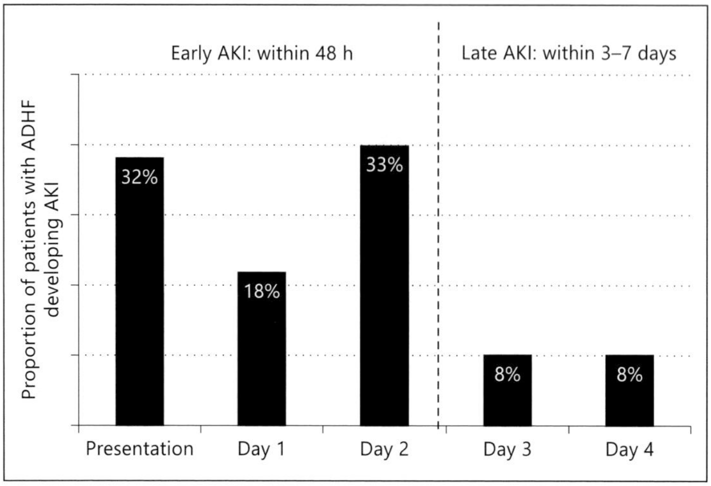

Fig. 4. Timing of acute kidney injury in the setting of acute decompensated heart failure. Adapted from Breidthardt et al. [50].

Ad Question 3: How do the timing, severity and duration relate to the mechanisms and outcomes of CRS type 1?

Recently, studies that had access to preadmission levels of serum creatinine revealed that AKI seems to be present already at presentation to the emergency department in about one third of consecutive patients with ADHF who have AKI at any time during hospitalization [50]. This may even be an underestimation which may arise due to the obvious dilemma of obtaining informed consent from the sickest patients.

Another 50% of AKI patients first fulfilled AKI criteria within the first 48 h in hospital, while only a minority developed late AKI after 48 h (fig. 4).

Given our incomplete understanding of the pathophysiology of CRS type 1, it may seem overzealous to differentiate predominate mechanisms according to the timing of AKI. However, we would like to highlight emerging hypothesis-generating observations of treatment effects: First, seemingly paradoxical, renal function often improves in response to standard therapy aimed to reduce filling pressures with loop diuretics and nitrates in warm and wet ADHF patients with AKI at presentation. Taken together, with the increasing recognition of increased renal venous congestion as a contributor to AKI in ADHF, this observation suggests that increased renal venous congestion may be the predominate mechanisms underlying AKI that is already present at presentation to the emergency department.

A post hoc analysis of Pre-RELAX found that drop in systolic blood pressure within the first 48 h of vasodilator therapy was an independent predictor of AKI by day 5 [51]. These data together with clinical experience support the concept that therapy-related reduction in renal perfusion pressure and/or ECFV are the predominate mechanisms underlying AKI that develops within the first days of hospitalization.

Usually, ACE inhibitors are slowly uptitrated during the course of hospitalization. The hemodynamic effect of ACE inhibitors (vasodilation of the vas efferens) and its associated drop in filtration pressure as well as 'pre-renal' kidney dysfunction due to the reduction in ECFV associated with intensified therapy with loop diuretics may be the predominate mechanism underlying AKI developing late during hospitalization.

Finally, severity and duration of AKI seem to relate to long-term patient outcome [52].

Consensus Recommendations for Future Research

- Confirm or rebut the hypothesis of the different 'AKI phenotypes' according to timing.
- Define the impact of timing of AKI on long-term outcome.

Consensus Recommendations for Clinical Practice

- Frequent (daily) monitoring of renal function during ADHF should complement clinical assessment in order to best guide therapy with diuretics and vasodilators with their profound impact on cardiorenal interactions.
- Daily assessment of limb perfusion (cold/warm extremities, recapillarization time, and others) and CVP as surrogates for renal filtration pressure and renal venous congestion as the two key hemodynamic variables determining renal function.

References

1 Goh CY, Vizzi G, De Cal M, Ronco C: Cardiorenal syndrome: a complex series of combined heart/kidney disorders. Contrib Nephrol. Basel, Karger, 2011, vol 174, pp 33–45.

2 Ronco C: Cardiorenal syndromes: definition and classification. Contrib Nephrol. Basel, Karger, 2010, vol 164, pp 33–38.

3 Eren Z, Ozveren O, Buvukoner E, Kaspar E, Degertekin M, Kantarci G: A single-centre study of acute cardiorenal syndrome: incidence, risk factors and consequences. Cardiorenal Med 2012;2:168–176.

4 Bagshaw SM, Cruz DN, Aspromonte N, Daliento L, Ronco F, Sheinfeld G, Anker SD, Anand I, Bellomo R, Berl T, Bobek I, Davenport A, Haapio M, Hillege H, House A, Katz N, Maisel A, Mankad S, McCullough P, Mebazaa A, Palazzuoli A, Ponikowski P, Shaw A, Soni S, Vescovo G, Zamperetti N, Zanco P, Ronco C, Acute Dialysis Quality Initiative Consensus Group: Epidemiology of cardiorenal syndromes: workgroup statements from the 7th ADQI Consensus Conference. Nephrol Dial Transplant 2010;25:1406–1416.

5 Kellum JA, Mehta RL, Levin A, Molitoris BA, Warnock DG, Shah SV, Joannidis M, Ronco C, Acute Kidney Injury Network (AKIN): Development of a clinical research agenda for acute kidney injury using an international, interdisciplinary, three-step modified Delphi process. Clin J Am Soc Nephrol 2008;3:887–894.
6 Hanada S, Takewa Y, Mizuno T, Tsukiyan T, Taenaka Y, Tatsumi E: Effect of the technique for assisting renal blood circulation on ischemic kidney in acute cardiorenal syndrome. J Artif Organs 2012;15:140–145.
7 Stevenson LW, Perloff JK: The limited reliability of physical signs for estimating hemodynamics in chronic heart failure. JAMA 1989;261:884–888.
8 Nohria A, Tsang SW, Fang JC, Lewis EF, Jarcho JA, Mudge GH, Stevenson LW: Clinical assessment identifies hemodynamic profiles that predict outcomes in patients admitted with heart failure. J Am Coll Cardiol 2003;41:1797–1804.
9 Mullens W, Abrahams Z, Francis GS, Sokos G, Taylor DO, Starling RC, Young JB, Tang WH: Importance of venous congestion for worsening of renal function in advanced decompensated heart failure. J Am Coll Cardiol 2009;53:589–596.
10 Damman K, van Deursen VM, Navis G, Voors AA, van Veldhuisen DJ, Hillege HL: Increased central venous pressure is associated with impaired renal function and mortality in a broad spectrum of patients with cardiovascular disease. J Am Coll Cardiol 2009;53:582–588.
11 Nohria A, Hasselblad V, Stebbins A, Pauly DF, Fonarow GC, Shah M, Yancy CW, Califf RM, Stevenson LW, Hill JA: Cardiorenal interactions: insights from the ESCAPE trial. J Am Coll Cardiol 2008;51:1268–1274.
12 Uthoff H, Breidthardt T, Klima T, Aschwanden M, Arenja N, Socrates T, Heinisch C, Noveanu M, Frischknecht B, Baumann U, Jaeger KA, Mueller C: Central venous pressure and impaired renal function in patients with acute heart failure. Eur J Heart Fail 2011;13:432–439.
13 Braam B, Cupples WA, Joles JA, Gaillard C: Systemic arterial and venous determinants of renal hemodynamics in congestive heart failure. Heart Fail Rev 2012;17:161–175.
14 Fiksen-Olsen MJ, Strick DM, Hawley H, Romero JC: Renal effects of angiotensin II inhibition during increases in renal venous pressure. Hypertension 1992;19:II137–II141.
15 DiBona GF: Nervous kidney. Interaction between renal sympathetic nerves and the renin-angiotensin system in the control of renal function. Hypertension 2000;36:1083–1088.
16 De Silva R, Loh H, Rigby AS, Nikitin NP, Witte KK, Goode K, Bhandari S, Nicholson A, Clark AL, Cleland JG: Epidemiology, associated factors, and prognostic outcomes of renal artery stenosis in chronic heart failure assessed by magnetic resonance angiography. Am J Cardiol 2007;100:273–279.
17 Machnik A, Neuhofer W, Jantsch J, Dahlmann A, Tammela T, Machura K, Park JK, Beck FX, Müller DN, Derer W, Goss J, Ziomber A, Dietsch P, Wagner H, van Rooijen N, Kurtz A, Hilgers KF, Alitalo K, Eckardt KU, Luft FC, Kerjaschki D, Titze J: Macrophages regulate salt-dependent volume and blood pressure by a vascular endothelial growth factor-C-dependent buffering mechanism. Nat Med 2009;15:545–552.
18 Virzì GM, Torregrossa R, Cruz DN, Chionh CY, de Cal M, Soni SS, Dominici M, Vescovo G, Rosner MH, Ronco C: Cardiorenal syndrome type 1 may be immunologically mediated: a pilot evaluation of monocyte apoptosis. Cardiorenal Med 2012;2:33–42.
19 Xu G, Ahn J, Chang S, Eguchi M, Ogier A, Han S, Park Y, Shim C, Jang Y, Yang B, Xu A, Wang Y, Sweeney G: Lipocalin-2 induces cardiomyocyte apoptosis by increasing intracellular iron accumulation. J Biol Chem 2012;287:4808–4817.
20 Ding L, Hanawa H, Ota Y, Hasegawa G, Hao K, Asami F, Watanabe R, Yoshida T, Toba K, Yoshida K, Ogura M, Kodama M, Aizawa Y: Lipocalin-2/neutrophil gelatinase-B associated lipocalin is strongly induced in hearts of rats with autoimmune myocarditis and in human myocarditis. Circ J 2010;74: 523–530.
21 Latouche C, El Moghrabi S, Messaoudi S, Nguyen Dinh Cat A, Hernandez-Diaz I, Alvarez de la Rosa D, Perret C, López Andrés N, Rossignol P, Zannad F, Farman N, Jaisser F: Neutrophil gelatinase-associated lipocalin is a novel mineralocorticoid target in the cardiovascular system. Hypertension 2012; 59:966–972.

22 Maisel A, Xue Y, Shah K, Mueller C, Nowak R, Peacock WF, Ponikowski P, Mockel M, Hogan C, Wu AH, Richards M, Clopton P, Filippatos GS, Di Somma S, Anand IS, Ng L, Daniels LB, Neath SX, Christenson R, Potocki M, McCord J, Terracciano G, Kremastinos D, Hartmann O, von Haehling S, Bergmann A, Morgenthaler NG, Anker SD: Increased 90-day mortality in patients with acute heart failure with elevated copeptin: secondary results from the Biomarkers in Acute Heart Failure (BACH) study. Circ Heart Fail 2011;4:613–620.
23 Shah RV, Truong QA, Gaggin HK, Pfannkuche J, Hartmann O, Januzzi JL Jr: Mid-regional pro-atrial natriuretic peptide and pro-adrenomedullin testing for the diagnostic and prognostic evaluation of patients with acute dyspnoea. Eur Heart J 2012;33: 2197–2205.
24 Maisel AS, Katz N, Hillege HL, Shaw A, Zanco P, Bellomo R, Anand I, Anker SD, Aspromonte N, Bagshaw SM, Berl T, Bobek I, Cruz DN, Daliento L, Davenport A, Haapio M, House AA, Mankad S, McCullough P, Mebazaa A, Palazzuoli A, Ponikowski P, Ronco F, Sheinfeld G, Soni S, Vescovo G, Zamperetti N, Ronco C, Acute Dialysis Quality Initiative Consensus Group: Biomarkers in kidney and heart disease. Nephrol Dial Transplant 2011;26:62–74.
25 Aldous SJ, Richards AM, Troughton R, Than M: ST2 has diagnostic and prognostic utility for all-cause mortality and heart failure in patients presenting to the emergency department with chest pain. J Card Fail 2012;18: 304–310.
26 Boisot S, Beede J, Isakson S, Chiu A, Clopton P, Januzzi J, Maisel AS, Fitzgerald RL: Serial sampling of ST2 predicts 90-day mortality following destabilized heart failure. J Card Fail 2008;14:732–738.
27 Ky B, French B, McCloskey K, Rame JE, McIntosh E, Shahi P, Dries DL, Tang WH, Wu AH, Fang JC, Boxer R, Sweitzer NK, Levy WC, Goldberg LR, Jessup M, Cappola TP: High-sensitivity ST2 for prediction of adverse outcomes in chronic heart failure. Circ Heart Fail 2011;4:180–187.
28 Dhillon OS, Narayan HK, Quinn PA, Squire IB, Davies JE, Ng LL: Interleukin-33 and ST2 in non-ST-elevation myocardial infarction: comparison with Global Registry of Acute Coronary Events Risk Scoring and NT-proBNP. Am Heart J 2011;161:1163–1170.
29 Tojo A, Gross SS, Zhang L, Tisher CC, Schmidt HH, Wilcox CS, Madsen KM: Immunocytochemical localization of distinct isoforms of nitric oxide synthase in the juxtaglomerular apparatus of normal rat kidney. J Am Soc Nephrol 1994;4:1438–1447.
30 Singh P, Okusa MD: The role of tubuloglomerular feedback in the pathogenesis of acute kidney injury. Contrib Nephrol. Basel, Karger, 2011, vol 174, pp 12–21.
31 Haase M, Kellum JA, Ronco C: Subclinical AKI – an emerging syndrome with important consequences. Nat Rev Nephrol 2012;8:735–739.
32 Campbell CY, Clarke W, Park H, Haq N, Barone BB, Brotman DJ: Usefulness of cystatin C and prognosis following admission for acute heart failure. Am J Cardiol 2009;104:389–392.
33 Lassus J, Harjola VP, Sund R, Siirilä-Waris K, Melin J, Peuhkurinen K, Pulkki K, Nieminen MS, for the FINN-AKVA Study Group: Prognostic value of cystatin C in acute heart failure in relation to other markers of renal function and NT-proBNP. Eur Heart J 2007;28:1841–1847.
34 Lassus JP, Nieminen MS, Peuhkurinen K, Pulkki K, Siirilä-Waris K, Sund R, Harjola VP, FINN-AKVA Study Group: Markers of renal function and acute kidney injury in acute heart failure: definitions and impact on outcomes of the cardiorenal syndrome. Eur Heart J 2010;31:2791–2798.
35 Shrestha K, Borowski AG, Troughton RW, Thomas JD, Klein AL, Tang WH: Renal dysfunction is a stronger determinant of systemic neutrophil gelatinase-associated lipocalin levels than myocardial dysfunction in systolic heart failure. J Card Fail 2011;17:472–478.
36 Cruz DN, Gaiao S, Maisel A, Ronco C, Devarajan P: Neutrophil gelatinase-associated lipocalin as a biomarker of cardiovascular disease: a systematic review. Clin Chem Lab Med 2012;50:1533–1545.
37 Alvelos M, Lourenço P, Dias C, Amorim M, Rema J, Leite AB, Guimarães JT, Almeida P, Bettencourt P: Prognostic value of neutrophil gelatinase-associated lipocalin in acute heart failure. Int J Cardiol (E-pub ahead of print).
38 Maisel AS, Mueller C, Fitzgerald R, Brikhan R, Hiestand BC, Iqbal N, Clopton P, van Veldhuisen DJ: Prognostic utility of plasma neutrophil gelatinase-associated lipocalin in patients with acute heart failure: the NGAL EvaLuation Along with B-type NaTriuretic Peptide in acutely decompensated heart failure (GALLANT) trial. Eur J Heart Fail 2011;13:846–851.

39 Alvelos M, Pimentel R, Pinho E, Gomes A, Lourenço P, Teles MJ, Almeida P, Guimarães JT, Bettencourt P: Neutrophil gelatinase-associated lipocalin in the diagnosis of type 1 cardiorenal syndrome in the general ward. Clin J Am Soc Nephrol 2011;6:476–481.
40 Aghel A, Shrestha K, Mullens W, Borowski A, Tang WH: Serum neutrophil gelatinase-associated lipocalin (NGAL) in predicting worsening renal function in acute decompensated heart failure. J Card Fail 2010;16:49–54.
41 Collins SP, Hart KW, Lindsell CJ, Fermann GJ, Weintraub NL, Miller KF, Roll SN, Sperling MI, Sawyer DB, Storrow AB: Elevated urinary neutrophil gelatinase-associated lipocalcin after acute heart failure treatment is associated with worsening renal function and adverse events. Eur J Heart Fail 2012;14:1020–1029.
42 Shrestha K, Shao Z, Singh D, Dupont M, Tang WH: Relation of systemic and urinary neutrophil gelatinase-associated lipocalin levels to different aspects of impaired renal function in patients with acute decompensated heart failure. Am J Cardiol 2012;110:1329–1335.
43 Dupont M, Shrestha K, Singh D, Awad A, Kovach C, Scarcipino M, Maroo AP, Tang WH: Lack of significant renal tubular injury despite acute kidney injury in acute decompensated heart failure. Eur J Heart Fail 2012;14:597–604.
44 Valente MA, Damman K, Dunselman PH, Hillege HL, Voors AA: Urinary proteins in heart failure. Prog Cardiovasc Dis 2012;55:44–55.
45 Di Somma S, De Berardinis B, Bongiovanni C, Marino R, Ferri E, Alfei B: Use of BNP and bioimpedance to drive therapy in heart failure patients. Congest Heart Fail 2010;16:S56–S61.
46 Parrinello G, Paterna S, Di Pasquale P, Torres D, Fatta A, Mezzero M, Scaglione R, Licata G: The usefulness of bioelectrical impedance analysis in differentiating dyspnea due to decompensated heart failure. J Card Fail 2008;14:676–686.
47 Valle R, Aspromonte N, Milani L, Peacock FW, Maisel AS, Santini M, Ronco C: Optimizing fluid management in patients with acute decompensated heart failure (ADHF): the emerging role of combined measurement of body hydration status and brain natriuretic peptide (BNP) levels. Heart Fail Rev 2011;16:519–529.
48 Milzman D, Napoli A, Hogan C, Zlidenny A, Janchar T: Thoracic impedance vs. chest radiograph to diagnose acute pulmonary edema in the ED. Am J Emerg Med 2009;27:770–775.
49 Whellan DJ, Droogan CJ, Fitzpatrick J, Adams S, McCarey MM, Andrel J, Mather P, Rubin S, Bonita R, Keith S: Change in intrathoracic impedance measures during acute decompensated heart failure admission: results from the Diagnostic Data for Discharge in Heart Failure Patients (3D-HF) Pilot Study. J Card Fail 2012;18:107–112.
50 Breidthardt T, Socrates T, Drexler B, Noveanu M, Heinisch C, Arenja N, Klima T, Züsli C, Reichlin T, Potocki M, Twerenbold R, Steiger J, Mueller C: Plasma neutrophil gelatinase-associated lipocalin for the prediction of acute kidney injury in acute heart failure. Crit Care 2012;16:R2.
51 Voors AA, Davison BA, Felker GM, Ponikowski P, Unemori E, Cotter G, Teerlink JR, Greenberg BH, Filippatos G, Teichman SL, Metra M, Pre-RELAX-AHF Study Group: Early drop in systolic blood pressure and worsening renal function in acute heart failure: renal results of Pre-RELAX-AHF. Eur J Heart Fail 2011;13:961–967.
52 Shirakabe A, Hata N, Kobayashi N, Shinada T, Tomita K, Tsurumi M, Matsushita M, Okazaki H, Yamamoto Y, Yokoyama S, Asai K, Mizuno K: Prognostic impact of acute kidney injury in patients with acute decompensated heart failure. Circ J 2013;77:687–696.
53 Acute Dialysis Quality Initiative (ADQI). www.ADQI.org (accessed January 10, 2013).

Michael Haase, MD
Department of Nephrology and Hypertension, Diabetes and Endocrinology
Otto von Guericke University Magdeburg
Leipziger Strasse 44, DE–39120 Magdeburg (Germany)
E-Mail michael.haase@charite.de

McCullough PA, Kellum JA, Mehta RL, Murray PT, Ronco C (eds): ADQI Consensus on AKI Biomarkers and Cardiorenal Syndromes. Contrib Nephrol. Basel, Karger, 2013, vol 182, pp 117–136
DOI: 10.1159/000349968

Pathophysiology of Cardiorenal Syndrome Type 2 in Stable Chronic Heart Failure: Workgroup Statements from the Eleventh Consensus Conference of the Acute Dialysis Quality Initiative (ADQI)

Dinna N. Cruz[a] · Andrew A. House[b] · Kai M. Schmidt-Ott[c] · Giorgio Vescovo[d] · John A. Kellum[e] · Claudio Ronco[f] · Peter A. McCullough[g] · for the Acute Dialysis Quality Initiative (ADQI) Consensus Group

[a]Division of Nephrology-Hypertension, Department of Medicine, University of California, San Diego, San Diego, Calif., USA; [b]Division of Nephrology, London Health Sciences Centre, London, Ont., Canada; [c]Department of Nephrology, Experimental and Clinical Research Center, Charité – Universitätsmedizin Berlin, and Max Delbrück Center for Molecular Medicine, Berlin, Germany; [d]Department of Internal Medicine, St. Bortolo Hospital, Vicenza, Italy; [e]Clinical Research, Investigation, and Systems Modeling of Acute Illness Center, Department of Critical Care Medicine, University of Pittsburgh, Pittsburgh, Pa., USA; [f]Department of Nephrology, International Renal Research Institute (IRRIV), San Bortolo Hospital, Vicenza, Italy; [g]St. John Providence Health System, Warren Mich., Providence Hospitals and Medical Centers, Southfield and Novi, Mich., St. John Macomb Oakland Center, Madison Heights, Mich., St. John Hospital and Medical Center, Detroit, Mich., USA

Abstract

In cardiorenal syndrome (CRS) type 2, chronic heart failure (HF) results in the onset or progression of chronic kidney disease (CKD). Examples of CRS type 2 (CRS2) include progressive CKD resulting from chronic HF in congenital or acquired heart disease or from repeated bouts of acute decompensated HF. Animal data and clinical studies indicate that extended periods of chronic HF result in altered renal hemodynamics followed by progressive renal pathology. Experimental and clinical data indicate that CRS2 is characterized by mild to moderate proteinuria, a progressive decline of glomerular filtration rate, and an elevated expression of renal injury biomarkers. Important pathophysiological triggers of renal disease progression include chronic increases in renal venous pressure, maladaptive activation of the renin-angiotensin-aldosterone axis and the sympathetic nervous system, as well as a chronic inflammatory state. Intrarenal oxidative stress and proinflammatory

ADQI 11 Workgroup members are listed in Appendix 2.

signaling precipitate structural injury, including glomerulosclerosis and tubulointerstitial fibrosis. Yet, clinical interventional trials that directly test the impact of renin-angiotensin system antagonists and β-blockers on the progression of CKD in CRS2 are lacking. Secondary analyses of trials designed to assess the impact of these agents on cardiovascular endpoints have failed to show a consistent benefit regarding renal functional parameters. In contrast, left ventricular assist device placement and cardiac resynchronization therapy in HF patients consistently improved renal function, suggesting a marked potential for reversibility in many cases of CRS2. Future research should be directed towards the evaluation of novel biomarkers to improve the diagnosis, severity grading as well as our understanding of the pathophysiology of CRS2. In addition, there is a need for interventional trials in HF patients to address long-term renal endpoints incorporating clinical information and measures of renal function as well as renal injury.

Cardiorenal syndrome (CRS) is defined as a complex pathophysiological disorder of the heart and kidneys in which acute or chronic dysfunction in one organ may induce acute or chronic dysfunction in the other organ. All subtypes are associated with increased mortality and morbidity, with a significant impact on health resource utilization. An understanding of the specific pathophysiologic mechanisms underlying each subtype will have important implications for risk factor modification, management, and future clinical trials.

CRS type 2 (CRS2) is characterized by chronic abnormalities in cardiac function leading to kidney injury or dysfunction. As such, the temporal relationship between the heart and kidney disease is an important aspect of the definition. While observational data clearly show that chronic heart and kidney disease commonly coexist, large database studies often assemble the patient cohort based on the existence of one disease process (e.g. chronic heart failure (HF)) and subsequently describe the prevalence of the other (chronic kidney disease (CKD)) [1, 2]. For example, CKD has been observed in 26–63% of CHF patients [1–3]. However, such studies are unable to determine which of the two disease processes was primary versus secondary, presenting challenges when attempting to classify patients into the CRS subtype definitions. In these situations, it has been suggested to use the term CRS 'type 2/4' [4].

CRS2 in Stable Chronic Heart Failure

In view of the above, the mere coexistence of cardiovascular disease and CKD is not sufficient to make a diagnosis of true CRS2. In the specific setting of stable CHF, we propose the following two prerequisites to make a diagnosis of CRS2; first, that CHF and CKD coexist in the patient, and second, CHF causally underlies the occurrence or progression of CKD. The latter should be supported by both tempo-

Chronic HF	Leading to	EITHER: New onset of CKD
Definition		Definition
Symptoms typical of HF Signs typical of HF (HF-REF) Reduced LVEF OR: (HF-PEF) Normal or mildly reduced EF and LV not dilated, with relevant structural disease and/or diastolic dysfunction (according to ESC, ACC/AHA)		Albuminuria and/or GFR <60 ml/min/1.73m^2 (according to KDIGO/KDOQI) **OR: Progression of CKD** Sustained ↓ eGFR of >5 ml/min/1.73m^2/year, or >10 ml/min/1.73m^2/5 years* OR: sustained increase in albuminuria

PLUS

Temporal association: A documented or presumed onset of congestive heart failure precedes the occurrence or progression of CKD

and

Pathophysiological plausibility: The manifestation and degree of kidney disease is plausibly explained by the underlying heart condition.

Fig. 1. Proposed definition of CRS2 in stable chronic HF. Reproduced with permission from ADQI [62].*National Collaborating Centre for Chronic Conditions. Chronic kidney disease: national clinical guideline for early identification and management in adults in primary and secondary care. London: Royal College of Physicians, September 2008.

ral association, i.e. documented or presumed onset of congestive HF temporally precedes the occurrence or progression of CKD, and by pathophysiological plausibility, that is, the manifestation and degree of kidney disease is plausibly explained by the underlying heart condition (fig. 1). One clear example of CRS2 would be congenital heart disease and 'cyanotic nephropathy', in which the heart disease temporally precedes any kidney disease. Another would be an acute cardiac event (e.g. acute myocardial infarction) resulting in chronic left ventricular (LV) dysfunction, then followed by the onset of CKD or progression of preexisting CKD.

Heterogeneity in the human HF population and aspects unique to cyanotic heart disease make it difficult to tease out specific mechanisms for CRS2 based on human studies. Therefore, animal studies provide important insight into the pathogenesis of CRS2 by mimicking human phenotypes (table 1). Although multiple and complex mechanisms have been proposed, the predominant mechanisms include neurohormonal activation, hemodynamic factors of renal hypoperfusion and venous congestion, inflammation and oxidative stress (fig. 2). Aside from specific mechanisms, we posit that multiple episodes of acute de-

Table 1. Animal models of heart failure and CRS2

Mechanism	Study	Model	Follow-up days	Biochemical outcomes	Histologic outcomes
RAAS activation	Onozato (2007)	Salt-sensitive Dahl rat HF model; eplerenone and/or trandolapril	126	ACEI or combination: lower creatinine, lower albuminuria Epl: lower albuminuria	ACEI: ↓ glomerular sclerosis, ↓ oxidative stress Epl or combination: ↓ glomerular sclerosis, ↓ oxidative stress and ↓ TGF-β
SNS activation	Rafiq (2012)	Rat aortic regurgitation; renal denervation, and/or olmesartan	180	Denervation lowers albumuria, plasma norepinephrine, renal Ang2 ARBlowersBNP,renalAng2	Denervation lessens glomerular injury and oxidative stress Denervation and ARB lessen further
Inflammation, fibrosis	Lu (2012)	Mouse coronary ligation; knockout for LOX-1	21	LOX-1 knockout had preserved GFR compared with control	LOX-1 knockout had less renal expression of pro-IL-1b, TGF-β, VCAM, fewer PMNs, ↓collagen/fibrosis
	Lekawanvijit (2012)	Rat coronary ligation	112	HF rats had lower GFR and more albuminuria than control	HF rats had marked increases in phosphorylated Smad2, KIM-1, IL-6, TGF-β and CD68+ cells, with significant interstitial fibrosis
	Entin-Meer (2012)	Rat coronary ligation	30	HF rats had lower GFR and more albuminuria than control	HF rats had marked increases in KIM-1, NGAL, CD3 and CD68+ cells, with significant interstitial fibrosis

compensation of the heart or kidney may contribute to the progression of HF and CKD (fig. 3). This is suggested by data showing that the number of previous hospitalizations for HF was a strong predictor of mortality after controlling for other known risk factors for HF survival [5]. Although repeat hospitalizations may be considered a marker of disease severity, suboptimal care or poor patient compliance, repeated acute decompensation could potentially lead to modification of cardiac structure, increased fibrosis and LV modeling. Moreover, on long-term follow-up of 70 patients with dilated cardiomyopathy, frequency of HF admissions was independently associated with development of CKD (defined as estimated creatinine clearance <60 ml/min) [6]. Similarly, the detrimental effect of acute kidney injury episodes on the development and progression of CKD has been demonstrated in animal models and in epidemiologic studies [7].

Animal Models and Pathophysiological Insights into CRS2

In experimental HF, there is a reduction in glomerular plasma flow, but an elevated glomerular filtration fraction due to efferent arteriolar vasoconstriction. When these changes persist for extended time periods (3–6 months in the rat

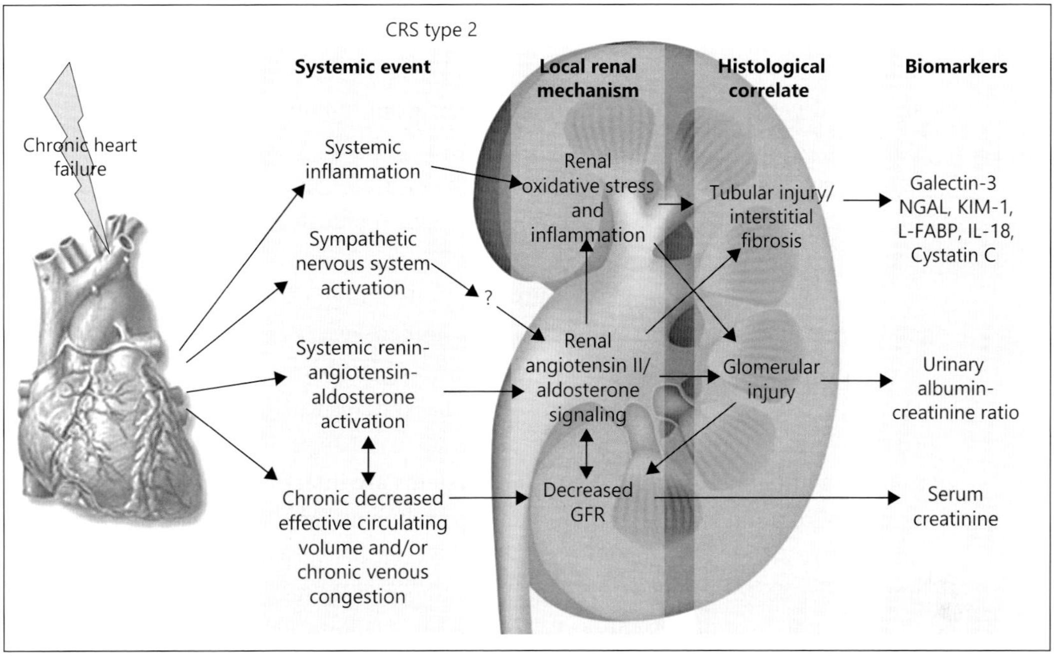

Fig. 2. Predominant pathophysiologic mechanisms of CRS2 in stable chronic HF. Reproduced with permission from ADQI [62]

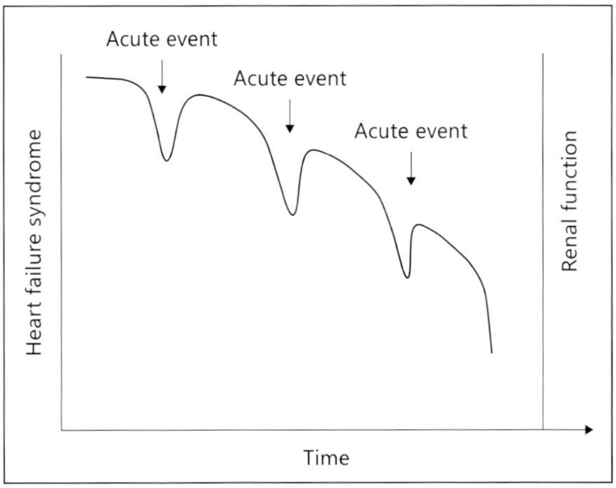

Fig. 3. Repeated acute events (HF decompensation and/or acute kidney injury) may contribute to the progression of chronic heart and kidney dysfunction. Reproduced with permission from ADQI [62].

model), there is increasing evidence of glomerular pathology as evidenced by albuminuria, podocyte injury and histological changes consistent with focal and segmental glomerulosclerosis. Variable decreases of glomerular filtration rate (GFR) have been reported in experimental chronic HF. Importantly, many of the changes appear to be related to systemic and local renal increases in sympathetic nervous system (SNS) and renin-angiotensin-aldosterone system (RAAS) activation.

One of the principle roles of a properly functioning cardiorenal axis is the maintenance of extracellular fluid volume homeostasis. A complex system of volume and pressure sensors, afferent and efferent feedback loops, local and distant vasoactive substances and neurohormonal systems with built-in redundancies serves to continuously monitor and adapt to changing extracellular fluid volume and blood pressure. When such systems are intact and working properly, they respond rapidly to ever-changing hemodynamics and volume status, and ensure adequate tissue perfusion and oxygen delivery. Some of the most important effector mechanisms include the SNS and RAAS.

In the setting of significant cardiac dysfunction, falling cardiac output and underfilling of the renal arterial tree results in activation of both SNS and RAAS [8]. It has long been recognized that the kidneys of patients with HF release substantial amounts of renin into the circulation [9], and this leads in turn to production of angiotensin II. Angiotensin II, working principally through the AT1 receptor, has broad-reaching effects. It is a powerful vasoconstrictor leading to increased systemic vascular resistance, venous tone and congestion, and has potent central nervous system effects on thirst, and activates the SNS. Angiotensin II also exerts significant effects on the kidneys stimulating tubular sodium reabsorption. Its powerful vasoconstrictive effects leads to preferential constriction of the efferent arteriole which in turn leads to increased glomerular filtration fraction, causing an increase in the oncotic pressure of the peritubular capillaries, enhancing the ability to return sodium and fluid to the circulation. The pivotal role of the RAAS system in this process has been demonstrated in animal models involving coronary artery ligation and subsequent infarction and HF [10] (table 1).

In support of clinical observational data implicating high venous pressure as an alternative and important cause of worsening GFR in HF, particularly with preserved ejection fraction and normal or high blood pressure, Kishimoto et al. [11] demonstrated in a dog model that renal venous hypertension, independent of changes in systemic arterial blood pressure, led not only to decreased renal blood flow and GFR, but also to increased renin release. This provides evidence for RAAS activation in the setting of HF characterized by venous hypertension and congestion, and not necessarily requiring decreased effective circulating volume as the stimulus.

In addition to the maladaptive worsening of pressure and volume overload, the chronic activation of the SNS and RAAS is thought to contribute to the progression of CKD characteristic of CRS2. In an elegant animal model of chronic volume overload, Rafiq et al. [12] surgically induced aortic regurgitation in uninephrectomized rats and examined intrarenal levels of norepinephrine and angiotensin II, as well as albuminuria, renal function and histologic evidence of podocyte injury and reactive oxygen species production. They found that chronic volume overload led to predictable changes in the structure and function of the heart, with corresponding increases in both intrarenal SNS and RAAS activity, and that progressive kidney injury could be prevented by renal denervation and angiotensin receptor blockade. The authors propose that activation of the SNS and local angiotensin II stimulates NADPH oxidase-dependent reactive oxygen species generation in the kidney, and these processes lead to podocyte injury and albuminuria.

Angiotensin II is also a potent stimulus for aldosterone release from the adrenal gland, leading to even further augmented sodium reabsorption in the distal nephron, contributing to pressure and volume overload. Moreover, aldosterone has been implicated in progression of CKD and renal fibrosis in a variety of clinical situations and through a number of mechanisms [13]. With an increased level of aldosterone in the kidney, pronounced oxidative stress occurs, via signaling from the paracrine glycoprotein galectin-3, pro-fibrotic cytokine transforming growth factor-β (TGF-β) is upregulated, followed by increases in fibronectin, leading to renal fibrosis and glomerulosclerosis. With respect to CRS2, Onozato et al. [14] examined Dahl salt-sensitive HF rats where increased aldosterone was felt to contribute to worsening kidney function through increased oxidant stress and production of TGF-β. In this experiment, untreated animals with HF developed subsequent proteinuria, elevated creatinine and glomerulosclerosis with increased expression of NADPH oxidase, TGF-β and fibronectin. They demonstrated that angiotensin-converting enzyme (ACE) inhibition and aldosterone blockade with eplerenone diminished oxidative stress, and TGF-β was more effectively suppressed by aldosterone blockade than by ACE inhibition. The combination was able to prevent histologic renal damage, and lowered both creatinine and proteinuria to control levels. It needs to be noted that this model did not separate the effects of hypertension from those of CHF on the progression of kidney disease. Yet, it suggests an interplay of hypertension-induced and HF-associated renal injury with a related and mutually perpetuating pathophysiology.

Inflammation is an additional non-hemodynamic mechanism postulated in the progression of CKD in the setting of CHF. Inflammatory responses are designed to provide protection and promote healing in disease states. Cardiac myocytes, under the stress of mechanical stretch or ischemia, are capable of producing a broad array of inflammatory cytokines and invoking elements of the innate immune response [15]. Furthermore, venous congestion may increase gut absorption of endotoxin leading to additional inflammatory responses – while venous congestion itself is a stimulus for peripheral synthesis and release of inflammatory mediators [16]. Clinical evidence for this pro-inflammatory state comes from observations that patients with severe HF have markedly elevated levels of tumor necrosis factor (TNF)-α, upregulation of soluble receptors for TNF, and a number of interleukins (IL) including IL-1β, IL-18 and IL-6, as well as several cellular adhesion molecules.

It is conceivable that these systemic responses to HF could contribute to distant organ damage in the kidneys, though direct evidence for this cardiorenal link is only now emerging. In an acute myocardial infarction model in mice, Lu et al. [17] induced significant suppression of LV function accompanied by a marked pro-oxidant and pro-inflammatory state. These animals demonstrated evidence of CRS type 1, with acute rise in creatinine, however kidney dysfunction did not return to baseline, and histologically the kidneys were found to express a variety of pro-inflammatory molecules, pro-interleukin-1b (IL-1b), vascular cell adhesion molecule-1 (VCAM-1), and TGF-β. Early changes included inflammatory cellular infiltrates, while animals sacrificed later in the course of study demonstrated perivascular, periglomerular and peritubular fibrosis with increased markers of collagen formation, consistent with the human phenotype of CRS2. Knockout mice with a deletion that prevents production of the pro-inflammatory molecule lectin-like oxidized low-density lipoprotein receptor-1 (LOX-1), demonstrated significant abrogation of these responses, with improved renal and cardiac function and morphology.

In a rat model of CRS2 [18], rats with underlying kidney dysfunction were subjected to left anterior descending coronary artery ligation, and further progression of advanced CKD ensued over 30 days of follow-up. HF rats demonstrated marked increases in biomarkers of renal injury neutrophil gelatinase-associated lipocalin (NGAL) and kidney injury molecule 1 (KIM-1), as well as patchy atrophic scarring with prominent clusters of CD3+ and CD68+ immune cells and significant interstitial fibrosis in renal tissue. In another rat model of CRS2, Lekawanvijit et al. [19] likewise performed left anterior descending coronary artery ligation to induce HF, and sacrificed animals after 16 weeks of chronic CHF. They also found increased expression of KIM-1 as well as the intracellular transcription factor phosphorylated Smad2, IL-6 and TGF-β in the HF animals. Associated with these inflammatory markers were infiltrates with CD68+ immune cells, and significant interstitial fibrosis. Similarly, in a rat model of right HF secondary to monocrotaline-induced pulmonary hypertension and characterized by venous congestion, Angelini et al. [20, 21] demonstrated at the same time neurohormonal activation with increased levels of BNP, angiotensin II and inflammation. Pro-inflammatory cytokines (IL-1α, β, IL-2, IL-4, IL-6, IL-10, TNF-α) were significantly increased and apoptosis was observed in the heart, lung, skeletal muscle, as well as in the vascular smooth muscle, tubular and glomerular cells of the kidney [Vescovo, pers. commun.]. In this model, NGAL detected by mRNA tissue extraction was also significantly increased in the HF rats compared to controls.

Clinical Studies and Pathophysiological Insights into CRS2

Data in humans provide support for these findings from animal studies. In a small autopsy study of 8 patients with CHF-related renal dysfunction, renal tissue also showed increased interstitial fibrosis, CD68+ immune cells and markers of oxidative stress (Rac1 expression and protein nitrosylation) [22]. Peritubular capillaries were dilated by 35%, believed to be indicative of increased venous pressure in the kidney. In other larger studies, increased central venous or right atrial pressures were associated with impaired renal function and independently related to all-cause mortality in a broad spectrum of patients with cardiovascular disease, including acute and chronic HF [23–25]. This relationship was more pronounced when reduced renal perfusion was also present [25]. As suggested by animal studies, the raised central venous pressure may be transmitted backwards to the renal veins raising renal interstitial pressure, and resulting in subsequent systemic and intra-renal activation of the RAAS and SNS.

Clinical trials in which specific pathophysiologic mechanisms were the therapeutic targets also serve as an important resource for understanding the mechanisms of CRS2. Unfortunately, large randomized clinical trials in CHF focus on cardiovascular mortality and adverse cardiac events, and renal outcomes are usually reported as safety endpoints, if at all. A limited number of studies reported on long-term changes in creatinine or GFR, micro- or macroalbuminuria or other aspects relevant to kidney damage such as markers of fibrosis or of inflammation, some as post hoc analyses (table 2).

Several studies have been carried out in chronic HF aimed to block the RAAS. These have been done both with ACE inhibitors (ACEIs) and angiotensin II receptor blockers (ARBs). The drugs most extensively studied have been enalapril and captopril for ACEIs and valsartan and candesartan for ARBs. In the SOLVD study of enalapril in CHF, the net deterioration of eGFR from baseline to 14 days was slightly greater in the enalapril group compared to

Table 2. Selected clinical trials on chronic heart failure and renal outcomes

Mechanism	Agent	Study	Study population	CKD patients excluded, yes/no
RAAS activation ACEIs/ARBS	Enalapril	Capes, 2000 SOLVD trial Testani, 2011, SOLVD	Clinically stable CHF (LVEF <35%) (n = 6,700)	no
		Ljungman, 1992 CONSENSUS	Heart failure NYHA IV (n = 250)	no
	Captopril	Hillege, 2003 CATS	Post-MI (n = 298)	no
	Valsartan	Anand, 2009 VALHEFT	Symptomatic heart failure EF <40% (n = 5,000)	yes Cr >2.5
	Candesartan	Jackson, 2009 CHARM-added	Heart failure NYHA II–IV EF <40% (n = 2,500)	no
SNS activation betablockers	Carvedilol	Wali, 2011 COPERNICUS CAPRICORN	Severe heart failure EF <25% (n = 4,200)	no
	Bisoprolol	Damman, 2011 Castagno, 2010 CIBIS II	CHF NYHA III–IV EF <35% (n = 2,660)	no, but substudy conducted in patients with Cr <3.4
	Nebivolol	Cohen-Solal, 2009 SENIORS	Symptomatic CHF Age >70 (n = 2,100)	yes, excluded if Cr >250 μmol/l
Aldosterone	Spironolactone	Pitt, 1999 Vardeny, 2012 RALES	CHF EF <35% (n = 1,600)	yes, excluded if Cr >2.5 mg/dl
		Zannad, 2001 RALES	Substudy on 261 patients	yes, excluded if Cr >2.5
	Eplerenone	Rossignol, 2012 EPHESUS	HF after MI EF <40% (n = 6,600)	no
		Zannad, 2011 EMPHASIS HF	HF NYHA II EF <35% (n = 2,700)	no
Renal hypoperfusion	Cardiac resynchronization	Boerrigter, 2008 MIRACLE	NYHA III–IV EF35% QRS >0.13 (n = 600)	no
	Left ventricle-assisted device	Sandner, 2009	Advanced CHF Bridge to Tx (n = 86)	no
Inflammation	Infliximab	Chung, 2003 ATTACH	CHF NYHA III–IV EF <35% (n = 150)	no
	Etanercept	Mann, 2004 RENEWAL	CHF NYHA III–IV EF <30% (n = 2,048)	no

Primary endpoint/ duration of follow-up	Renal outcomes Cr/eGFR	Albuminuria
Composite of CV death and re-admission for HF/48 months	Not stated	Decreased proteinuria in diabetic patients, not in non-diabetics
	14 days decline GFR >20% exited in 20% increase mortality	Not stated
Crude mortality 6 months	Average initial increase Cr 10–15%	Not stated
Echo ventricular remodeling 1 year	Decline GFR 5.5 in placebo and 0.5 in captopril	Not stated
Mortality + first morbid event 1,000 days	Slight decrease in GFR in valsartan (–3.9 vs. placebo)	Proteinuria associated with 28% increased mortality
CV death Hospital admission 41 months	Not stated	62% increased death in patients with microalbuminuria and 76% with macroalbuminuria. Candesartan did not prevent excessive excretion of urinary proteins
All-cause mortality and hospitalization 10 months	Transient increase in serum creatinine more common with carvedilol (4.6 vs. 1.8% in placebo, $p < 0.001$) among CKD patients	Not stated
All-cause mortality hospitalization 18 months	Not stated	Not stated
All-cause mortality hospitalization 30 months	No difference in ΔeGFR between nebivolol and placebo	Not stated
Death of any cause 36 months	WRF 17% in spironolactone vs. 7% placebo but risk of death maintained Cr increase 0.05 to 0.1 after 1 year follow-up	Not stated
Death of any cause 36 months	Not stated	Not stated
	Other: markers of collagen synthesis (PINP PICP PIIINP) decreased with spironolactone	
All deaths hospitalization 24 months	Cr increase 0.06 after 1 year 4.6 after 2 years vs. placebo 2.7	Not stated
All deaths hospitalization 21 months	Cr increase 0.09 at trial cut-off	Not stated
Improved, unchanged, worsened symptom score 12 months	GFR increased 2.7 in patients with GFR 30–60	Not stated
Renal function 6 months	Renal function improves after LVAD implantation	Not stated
Improvement clinical status (NYHA). Death, hospitalization 28 weeks	Not stated	Not stated
Clinical status at 24 weeks. Death, hospitalization	Not stated	Not stated

placebo. While early worsening of renal function was associated with increased mortality in the placebo group, it was free from adverse prognostic significance in the enalapril group [26]. In the same study, diabetic patients showed a decreased proteinuria with enalapril treatment [27]. An additional multivariable analysis suggested that despite a higher incidence of early worsening renal function in the enalapril group, there was no risk of longer term deterioration of eGFR compared with placebo [28]. In the CONSENSUS trial, patients randomized to enalapril showed an average increase in creatinine of 10–15% [29]. Hillege et al. [30] in post-MI patients with HF, although they observed a general deterioration of renal function after the acute event, reported that captopril preserved GFR when compared to placebo. In the VALHEFT trial a slight decrease in GFR in valsartan patients was observed (–3.9 vs. placebo). The presence of dipstick proteinuria was associated with a 28% increase in mortality risk [31]. In the CHARM-Added trial with candesartan, a 62 and 76% increased risk of death was observed in patients with micro- and macroalbuminuria, respectively. Candesartan did not prevent proteinuria [32]. Together, the current state of published evidence remains inconclusive regarding the role of ACEIs or ARBs in the prevention of the progression of CRS2.

CHF treatment with direct inhibitors of aldosterone receptors brought about a significant improvement in terms of survival and hospitalizations. Available data on spironolactone in the RALES trial showed worsening renal function in 17% of treated patients versus 7% in the placebo group, but the benefits on mortality were present in spite of this. Creatinine increased by approximately 0.05–0.1 mg/dl after 1 year follow-up in the spironolactone group with no significant change in the placebo group [33]. In the EPHESUS trial with eplerenone in post-MI patients with CHF, a creatinine increase of 0.06 mg/dl was observed after 1 year and 4.6 mg/dl after 2 years, while the placebo group had only an increase of 2.7 mg/dl. In the EPHESUS trial, eplerenone produced a 0.09 mg/dl increase in creatinine at trial cut-off [34–36]. A decrease in surrogate markers of collagen synthesis (PINP, PICP, PIIINP) after spironolactone treatment suggests that fibrosis could be reduced after anti-aldosterone treatment [37].

It is important to note that a slight, expected, increase in creatinine, particularly in trials with inhibitors of RAAS, does not necessarily mean a progression of the CRS2. Data on survival, symptoms, exercise capacity, proteinuria or albuminuria are clearly visible with these drugs despite a small increase in serum creatinine. It is likely that serum creatinine, as a marker for GFR, is an inadequate biomarker for actual progression of disease given the difficulty in differentiating true progressive CKD from hemodynamic (and therefore potentially reversible) changes in GFR due to RAAS blockade and effects on filtration fraction.

β-Blockers have become first-line treatment for CHF after the publication of COPERNICUS, CAPRICORN and CIBIS trials. In a meta-analysis of the CAPRICORN and COPERNICUS studies, carvedilol was associated with an increased relative incidence in transient increase in serum creatinine without need for dialysis (4.6 vs. 1.8% in placebo, $p < 0.001$) among CKD patients [38]. In the CIBIS trial, GFR was shown to be lower in patients with congestion. Bisoprolol showed greater benefits in patients with CKD (GFR <45) compared to placebo [39] and in this trial β-blocker treatment did not show any increase in serum creatinine. In elderly patients, nebivolol treatment did not modify GFR [40]. Overall there is inconsistency in the data on impact of β-blockers on creatinine in these trials.

Inflammation has been shown to play an important role in the pathophysiology of HF both in terms of progression of cardiac damage and deterioration of the clinical status. Inflammation also plays an important role in the pathogenesis of kidney damage and progression of deterioration of renal function. This observation led several investigators to use TNF-α inhibitors in the treatment of HF. The results of several trials with infliximab or etanercept have been rather disappointing. Indeed either futility or excessive mortality have been shown with these drugs, despite an observed decrease in plasma levels of hsCRP and IL-6. This is corroborated by the results of the ATTACH, RECOVER and RENASSAINCE (RENEWAL) trials [41, 42]. However there are no reports in these trials of renal function.

Hypoperfusion in HF can be improved with cardiac resynchronization or left ventricular assist devices. In one study, cardiac resynchronization increased GFR by 2.7 ml/min in the subgroup of patients with GFR between 30 and 60 [43]. Sandner et al. [44] showed that renal function improves after left ventricular assist device implantation for bridging to cardiac transplantation. In an observational study on 10 children with cyanotic congenital heart disease, urinary excretion of albumin and markers of tubular injury, such as brush-border leucine-aminopeptidase and lysosomal N-acetyl-β-D-glucosaminidase (NAG) decreased significantly after palliative heart surgery. This was paralleled by an improvement in oxygen saturation levels and a decrease in hematocrit to near-normal levels [45].

Biomarkers of Renal Function

To date, assessment of renal dysfunction in clinical HF studies has been largely limited to traditional biomarkers such as creatinine (or its derivative, eGFR) and urine protein and albumin excretion. In CHF, both impaired renal func-

Table 3. Selected studies on novel renal biomarkers in chronic heart failure

Study	Study population	Specimen	Biomarker	HF vs. control subjects
Tang, 2008	Clinically stable CHF outpatients (n = 139)	Plasma	Cystatin C	No controls
Damman, 2008, 2010	Clinically stable CHF outpatients (n = 90)	Urine	NGAL NAG KIM-1	HF > controls HF > controls HF > controls
Jungbauer, 2011	CHF (n = 173)	Urine	NGAL NAG KIM-1	HF = controls HF = controls HF > controls
Poniatowski, 2009	CHF due to CAD (n = 150)	Serum Urine	NGAL Cystatin C NGAL Cystatin C	No controls
Bolignano, 2009	CHF (elderly, age >65 years) (n = 46)	Serum	NGAL	HF > controls
Yndestad, 2009	CHF (n = 150)	Serum	NGAL	HF > controls
Shrestha, 2011	CHF (LVEF ≤35%) (n = 130)	Plasma	NGAL	No controls

CAD = Coronary artery disease; CHF = chronic heart failure; Cr = creatinine; GFR = glomerular filtration rate; HF = heart failure; NS = not stated.

tion (as indicated by elevated creatinine or decreased eGFR) and increased urinary albumin excretion are powerful and independent predictors of prognosis [32, 46]. Both are associated with heightened risk for death, cardiovascular death, and hospitalization, in patients with both preserved as well as reduced LVEF. In CKD, the level of eGFR and albuminuria has been shown to be prognostic of long-term renal outcomes [47, 48]; this has not been demonstrated in CHF [49]. More recently, novel renal biomarkers, such as cystatin C (CysC), NGAL, KIM-1, and NAG have also been studied in CHF patients [50–57] (table 3). In many, but not all, studies, levels of these biomarkers are modestly elevated in CHF compared to control subjects, even among those with apparently normal renal function. While some of these biomarkers appear to have some prognostic properties with regard to adverse cardiovascular events, no study has looked at long-term change in renal function as an outcome.

Plasma CysC, believed to be a more sensitive marker of impaired GFR compared to creatinine, has been reported as an independent predictor of death,

Biomarker correlates with (in HF patients only)			Renal outcomes
renal function	clinical HF severity	natriuretic peptides	
Yes	Yes	Yes	NS
No	NS	No	NS
Yes	NS	Yes	NS
No	NS	Yes	NS
Yes, with albuminuria No, with Cr/eGFR	No	No	No association with diuretic dose
Yes	Yes	Yes	Elevated in patients on higher (vs. lower) diuretic doses
Yes, with albuminuria No, with Cr/eGFR	Yes	Yes	Elevated in patients on higher (vs. lower) diuretic doses
Yes	Yes	NS	NS
Yes	NS	NS	NS
Yes	Yes	NS	NS
NS	NS	NS	NS
No	Yes	NS	NS
NS	Yes	Yes	NS
Yes	Yes	Yes	

cardiac transplantation and HF hospitalizations [52, 54]. CysC also correlated with NT-pro-BNP levels and measures of LV dysfunction [52]. Studies assessing urinary CysC levels in CHF populations are currently lacking.

Recently, a systematic review summarized studies on NGAL in a spectrum of cardiovascular diseases, including CHF [50]. Animal and human tissue studies demonstrated that NGAL is highly expressed in failing myocardium and myocarditis, and it is also expressed in atherosclerotic plaques. Data on NGAL in a CRS2 rat model have been also been discussed above [18]. In clinical studies, blood and urine NGAL levels generally correlate with creatinine or eGFR; they also correlate with clinical and biochemical markers (e.g. natriuretic peptides) of HF severity in some, but not all, studies [50]. Systemic NGAL levels have also been associated with increased mortality and HF hospitalization [50, 53, 55, 58, 59].

Analogous findings have also been reported for NAG and KIM-1. Urine levels of KIM-1, but not NAG or NGAL, are elevated in symptomatic HF compared to controls [60]. KIM-1 and NAG levels correlated with severity of

HF, and were also predictors of all-cause mortality and HF hospitalization on survival curve analysis. In 2,130 patients participating in the GISSI-HF trial, CHF patients who experienced a composite outcome of death or HF hospitalization had a significantly lower eGFR and a significantly higher urinary excretion of albumin, NAG, KIM-1, and NGAL [53]. On multivariable regression, the strongest association was seen with urinary NAG.

Although data on novel renal biomarkers in CHF is limited to date, they have great potential to enhance our understanding of CRS2 pathophysiology and should be evaluated in future large randomized controlled trials in HF. Indeed, the MinerAlocorticoid Receptor Antagonist Tolerability Study (ARTS) will examine the effects of the non-steroidal, mineralocorticoid receptor antagonist BAY 94-8862 on biomarkers of cardiac and renal function and injury, including the natriuretic peptides, KIM-1, NGAL, and CysC [61]. Results from this small study will undoubtedly inform future CRS2 trials.

Conclusion

In CRS2, the HF temporally precedes the occurrence or progression of CKD, and the manifestation and degree of kidney disease is plausibly explained by the underlying heart condition. Multiple and complex mechanisms have been proposed for CRS2; the most predominant being systemic and local renal increases in SNS and RAAS activation, hemodynamic factors of renal hypoperfusion and venous congestion, and inflammatory pathways. Animal models provide important insight into the pathogenesis of CRS2 by mimicking human phenotypes. Of the multitude of randomized controlled trials in CHF, relatively few reported long-term renal outcomes. While increased risk of kidney functional decline has been reported in some studies of RAAS inhibitors, most of the difference seen at the end of follow-up appeared to be attributable to the early treatment period. To date, renal outcomes have been largely limited to traditional biomarkers such as creatinine, eGFR, and urinary albumin excretion. Novel renal biomarkers appear to have some prognostic properties with regard to adverse cardiovascular events in CHF patients, but no study has looked at long-term change in renal function as an outcome. Ongoing and future interventional trials incorporating biomarkers of cardiac and renal function and injury among the endpoints will enhance our understanding of the pathophysiologic mechanisms behind CRS2.

References

1 Heywood JT, Fonarow GC, Costanzo MR, Mathur VS, Wigneswaran JR, Wynne J: High prevalence of renal dysfunction and its impact on outcome in 118,465 patients hospitalized with acute decompensated heart failure: a report from the ADHERE database. J Card Fail 2007;13:422–430.

2 Hebert K, Dias A, Delgado MC, Franco E, Tamariz L, Steen D, Trahan P, Major B, Arcement LM: Epidemiology and survival of the five stages of chronic kidney disease in a systolic heart failure population. Eur J Heart Fail 2010;12:861–865.

3 Cruz DN, Bagshaw SM: Heart-kidney interaction: epidemiology of cardiorenal syndromes. Int J Nephrol 2010;2011:351291.

4 Bagshaw SM, Cruz DN, Aspromonte N, Daliento L, Ronco F, Sheinfeld G, Anker SD, Anand I, Bellomo R, Berl T, et al: Epidemiology of cardio-renal syndromes: workgroup statements from the 7th ADQI Consensus Conference. Nephrol Dial Transplant 2010; 25:1406–1416.

5 Setoguchi S, Stevenson LW, Schneeweiss S: Repeated hospitalizations predict mortality in the community population with heart failure. Am Heart J 2007;154:260–266.

6 Tanaka K, Ito M, Kodama M, Maruyama H, Hoyano M, Mitsuma W, Iino N, Hirono S, Okura Y, Gejyo F, et al: Longitudinal change in renal function in patients with idiopathic dilated cardiomyopathy without renal insufficiency at initial diagnosis. Circ J 2007; 71:1927–1931.

7 Chawla LS, Kimmel PL: Acute kidney injury and chronic kidney disease: an integrated clinical syndrome. Kidney Int 2012;82:516–524.

8 Bongartz LG, Cramer MJ, Doevendans PA, Joles JA, Braam B: The severe cardiorenal syndrome: 'Guyton revisited'. Eur Heart J 2005;26:11–17.

9 Merrill AJ, Morrison JL, Branno ES: Concentration of renin in renal venous blood in patients with chronic heart failure. Am J Med 1946;1:468.

10 Ichikawa I, Pfeffer JM, Pfeffer MA, Hostetter TH, Brenner BM: Role of angiotensin II in the altered renal function of congestive heart failure. Circ Res 1984;55:669–675.

11 Kishimoto T, Maekawa M, Abe Y, Yamamoto K: Intrarenal distribution of blood flow and renin release during renal venous pressure elevation. Kidney Int 1973;4:259–266.

12 Rafiq K, Noma T, Fujisawa Y, Ishihara Y, Arai Y, Nabi AH, Suzuki F, Nagai Y, Nakano D, Hitomi H, et al: Renal sympathetic denervation suppresses de novo podocyte injury and albuminuria in rats with aortic regurgitation. Circulation 2012;125:1402–1413.

13 Remuzzi G, Cattaneo D, Perico N: The aggravating mechanisms of aldosterone on kidney fibrosis. J Am Soc Nephrol 2008;19: 1459–1462.

14 Onozato ML, Tojo A, Kobayashi N, Goto A, Matsuoka H, Fujita T: Dual blockade of aldosterone and angiotensin II additively suppresses TGF-β and NADPH oxidase in the hypertensive kidney. Nephrol Dial Transplant 2007;22:1314–1322.

15 Colombo PC, Ganda A, Lin J, Onat D, Harxhi A, Iyasere JE, Uriel N, Cotter G: Inflammatory activation: cardiac, renal, and cardio-renal interactions in patients with the cardiorenal syndrome. Heart Fail Rev 2012; 17:177–190.

16 Colombo PC, Onat D, Sabbah HN: Acute heart failure as 'acute endothelitis'–interaction of fluid overload and endothelial dysfunction. Eur J Heart Fail 2008;10:170–175.

17 Lu J, Wang X, Wang W, Muniyappa H, Deshmukh A, Hu C, Das K, Mehta JL: Abrogation of lectin-like oxidized LDL receptor-1 attenuates acute myocardial ischemia-induced renal dysfunction by modulating systemic and local inflammation. Kidney Int 2012;82:436–444.

18 Entin-Meer M, Ben-Shoshan J, Maysel-Auslender S, Levy R, Goryainov P, Schwartz I, Barshack I, Avivi C, Sharir R, Keren G: Accelerated renal fibrosis in cardiorenal syndrome is associated with long-term increase in urine neutrophil gelatinase-associated lipocalin levels. Am J Nephrol 2012;36:190–200.

19 Lekawanvijit S, Kompa AR, Zhang Y, Wang BH, Kelly DJ, Krum H: Myocardial infarction impairs renal function, induces renal interstitial fibrosis, and increases renal KIM-1 expression: implications for cardiorenal syndrome. Am J Physiol Heart Circ Physiol 2012;302:H1884–H1893.

20 Angelini A, Castellani C, Ravara B, Franzin C, Pozzobon M, Tavano R, Libera LD, Papini E, Vettor R, De Coppi P, et al: Stem-cell therapy in an experimental model of pulmonary hypertension and right heart failure: role of paracrine and neurohormonal milieu in the remodeling process. J Heart Lung Transplant 2011;30:1281–1293.
21 Dalla Libera L, Ravara B, Angelini A, Rossini K, Sandri M, Thiene G, Battista Ambrosio G, Vescovo G: Beneficial effects on skeletal muscle of the angiotensin II type 1 receptor blocker irbesartan in experimental heart failure. Circulation 2001;103:2195–2200.
22 Tanaka M, Yoshida H, Furuhashi M, Togashi N, Koyama M, Yamamoto S, Yamashita T, Okazaki Y, Ishimura S, Ota H, et al: Deterioration of renal function by chronic heart failure is associated with congestion and oxidative stress in the tubulointerstitium. Intern Med 2011;50:2877–2887.
23 Damman K, van Deursen VM, Navis G, Voors AA, van Veldhuisen DJ, Hillege HL: Increased central venous pressure is associated with impaired renal function and mortality in a broad spectrum of patients with cardiovascular disease. J Am Coll Cardiol 2009;53:582–588.
24 Mullens W, Abrahams Z, Francis GS, Sokos G, Taylor DO, Starling RC, Young JB, Tang WHW: Importance of venous congestion for worsening of renal function in advanced decompensated heart failure. J Am Coll Cardiol 2009;53:589–596.
25 Damman K, Navis G, Smilde TD, Voors AA, van der Bij W, van Veldhuisen DJ, Hillege HL: Decreased cardiac output, venous congestion and the association with renal impairment in patients with cardiac dysfunction. Eur J Heart Fail 2007;9:872–878.
26 Testani JM, Kimmel SE, Dries DL, Coca SG: Prognostic importance of early worsening renal function after initiation of angiotensin-converting enzyme inhibitor therapy in patients with cardiac dysfunction. Circ Heart Fail 2011;4:685–691.
27 Capes SE, Gerstein HC, Negassa A, Yusuf S: Enalapril prevents clinical proteinuria in diabetic patients with low ejection fraction. Diabetes Care 2000;23:377–380.
28 Bansal N, Tighiouart H, Weiner D, Griffith J, Vlagopoulos P, Salem D, Levin A, Sarnak MJ: Anemia as a risk factor for kidney function decline in individuals with heart failure. Am J Cardiol 2007;99:1137–1142.
29 Ljungman S, Kjekshus J, Swedberg K: Renal function in severe congestive heart failure during treatment with enalapril – the Cooperative North Scandinavian Enalapril Survival Study (CONSENSUS) Trial. Am J Cardiol 1992;70:479–487.
30 Hillege HL, van Gilst WH, van Veldhuisen DJ, Navis G, Grobbee DE, de Graeff PA, de Zeeuw D: Accelerated decline and prognostic impact of renal function after myocardial infarction and the benefits of ACE inhibition: the CATS randomized trial. Eur Heart J 2003; 24:412–420.
31 Anand IS, Bishu K, Rector TS, Ishani A, Kuskowski MA, Cohn JN: Proteinuria, chronic kidney disease, and the effect of an angiotensin receptor blocker in addition to an angiotensin-converting enzyme inhibitor in patients with moderate to severe heart failure. Circulation 2009;120:1577–1584.
32 Jackson CE, Solomon SD, Gerstein HC, Zetterstrand S, Olofsson B, Michelson EL, Granger CB, Swedberg K, Pfeffer MA, Yusuf S, et al: Albuminuria in chronic heart failure: prevalence and prognostic importance. Lancet 2009;374:543–550.
33 Pitt B, Zannad F, Remme WJ, Cody R, Castaigne A, Perez A, Palensky J, Wittes J: The effect of spironolactone on morbidity and mortality in patients with severe heart failure. Randomized Aldactone Evaluation Study Investigators. N Engl J Med 1999;341: 709–717.
34 Rossignol P, Cleland JG, Bhandari S, Tala S, Gustafsson F, Fay R, Lamiral Z, Dobre D, Pitt B, Zannad F: Determinants and consequences of renal function variations with aldosterone blocker therapy in heart failure patients after myocardial infarction: insights from the Eplerenone Post-Acute Myocardial Infarction Heart Failure Efficacy and Survival Study. Circulation 2012;125:271–279.
35 Zannad F, McMurray JJ, Krum H, van Veldhuisen DJ, Swedberg K, Shi H, Vincent J, Pocock SJ, Pitt B: Eplerenone in patients with systolic heart failure and mild symptoms. N Engl J Med 2011;364:11–21.

36 Zannad F, Gattis Stough W, Rossignol P, Bauersachs J, McMurray JJ, Swedberg K, Struthers AD, Voors AA, Ruilope LM, Bakris GL, et al: Mineralocorticoid receptor antagonists for heart failure with reduced ejection fraction: integrating evidence into clinical practice. Eur Heart J 2012;33:2782–2795.
37 Zannad F, Dousset B, Alla F: Treatment of congestive heart failure: interfering the aldosterone-cardiac extracellular matrix relationship. Hypertension 2001;38:1227–1232.
38 Wali RK, Iyengar M, Beck GJ, Chartyan DM, Chonchol M, Lukas MA, Cooper C, Himmelfarb J, Weir MR, Berl T, et al: Efficacy and safety of carvedilol in treatment of heart failure with chronic kidney disease: a meta-analysis of randomized trials. Circulation Heart failure 2011;4:18–26.
39 Castagno D, Jhund PS, McMurray JJ, Lewsey JD, Erdmann E, Zannad F, Remme WJ, Lopez-Sendon JL, Lechat P, Follath F, et al: Improved survival with bisoprolol in patients with heart failure and renal impairment: an analysis of the cardiac insufficiency bisoprolol study II (CIBIS-II) trial. Eur J Heart Fail 2010;12:607–616.
40 Cohen-Solal A, Kotecha D, van Veldhuisen DJ, Babalis D, Bohm M, Coats AJ, Roughton M, Poole-Wilson P, Tavazzi L, Flather M: Efficacy and safety of nebivolol in elderly heart failure patients with impaired renal function: insights from the SENIORS trial. Eur J Heart Fail 2009;11:872–880.
41 Chung ES, Packer M, Lo KH, Fasanmade AA, Willerson JT: Randomized, double-blind, placebo-controlled, pilot trial of infliximab, a chimeric monoclonal antibody to tumor necrosis factor-α, in patients with moderate-to-severe heart failure: results of the anti-TNF Therapy Against Congestive Heart Failure (ATTACH) trial. Circulation 2003;107:3133–3140.
42 Mann DL, McMurray JJ, Packer M, Swedberg K, Borer JS, Colucci WS, Djian J, Drexler H, Feldman A, Kober L, et al: Targeted anticytokine therapy in patients with chronic heart failure: results of the Randomized Etanercept Worldwide Evaluation (RENEWAL). Circulation 2004;109:1594–1602.
43 Boerrigter G, Costello-Boerrigter LC, Abraham WT, Sutton MG, Heublein DM, Kruger KM, Hill MR, McCullough PA, Burnett JC Jr: Cardiac resynchronization therapy improves renal function in human heart failure with reduced glomerular filtration rate. J Card Fail 2008;14:539–546.
44 Sandner SE, Zimpfer D, Zrunek P, Rajek A, Schima H, Dunkler D, Grimm M, Wolner E, Wieselthaler GM: Renal function and outcome after continuous flow left ventricular assist device implantation. Ann Thorac Surg 2009;87:1072–1078.
45 Awad H, el-Safty I, Abdel-Gawad M, el-Said S: Glomerular and tubular dysfunction in children with congenital cyanotic heart disease: effect of palliative surgery. Am J Med Sci 2003;325:110–114.
46 Hillege HL, Nitsch D, Pfeffer MA, Swedberg K, McMurray JJ, Yusuf S, Granger CB, Michelson EL, Ostergren J, Cornel JH, et al: Renal function as a predictor of outcome in a broad spectrum of patients with heart failure. Circulation 2006;113:671–678.
47 Halbesma N, Jansen DF, Heymans MW, Stolk RP, de Jong PE, Gansevoort RT: Development and validation of a general population renal risk score. Clin J Am Soc Nephrol 2011;6:1731–1738.
48 Tangri N, Stevens LA, Griffith J, Tighiouart H, Djurdjev O, Naimark D, Levin A, Levey AS: A predictive model for progression of chronic kidney disease to kidney failure. JAMA 2011;305:1553–1559.
49 Khan NA, Ma I, Thompson CR, Humphries K, Salem DN, Sarnak MJ, Levin A: Kidney function and mortality among patients with left ventricular systolic dysfunction. J Am Soc Nephrol 2006;17:244–253.
50 Cruz DN, Gaiao S, Maisel A, Ronco C, Devarajan P: Neutrophil gelatinase-associated lipocalin as a biomarker of cardiovascular disease: a systematic review. Clin Chem Lab Med 2012;50:1533–1545.
51 Cruz DN, Fard A, Clementi A, Ronco C, Maisel A: Role of biomarkers in the diagnosis and management of cardiorenal syndromes. Semin Nephrol 2012;32:79–92.
52 Tang WH, Van Lente F, Shrestha K, Troughton RW, Francis GS, Tong W, Martin MG, Borowski AG, Jasper S, Starling RC, et al: Impact of myocardial function on cystatin C measurements in chronic systolic heart failure. J Card Fail 2008;14:394–399.

53 Damman K, Masson S, Hillege HL, Maggioni AP, Voors AA, Opasich C, van Veldhuisen DJ, Montagna L, Cosmi F, Tognoni G, et al: Clinical outcome of renal tubular damage in chronic heart failure. Eur Heart J 2011;32: 2705–2712.
54 Damman K, van der Harst P, Smilde TD, Voors AA, Navis G, van Veldhuisen DJ, Hillege HL: Use of cystatin C levels in estimating renal function and prognosis in patients with chronic systolic heart failure. Heart 2012;98:319–324.
55 Damman K, van Veldhuisen DJ, Navis G, Vaidya VS, Smilde TD, Westenbrink BD, Bonventre JV, Voors AA, Hillege HL: Tubular damage in chronic systolic heart failure is associated with reduced survival independent of glomerular filtration rate. Heart 2010;96:1297–1302.
56 Damman K, van Veldhuisen DJ, Navis G, Voors AA, Hillege HL: Urinary neutrophil gelatinase associated lipocalin, a marker of tubular damage, is increased in patients with chronic heart failure. Eur J Heart Fail 2008; 10:997–1000.
57 Nymo SH, Ueland T, Askevold ET, Flo TH, Kjekshus J, Hulthe J, Wikstrand J, McMurray J, van Veldhuisen DJ, Gullestad L, et al: The association between neutrophil gelatinase-associated lipocalin and clinical outcome in chronic heart failure: results from CORONA. J Intern Med 2012;271:436–443.
58 Bolignano D, Basile G, Parisi P, Coppolino G, Nicocia G, Buemi M: Increased plasma neutrophil gelatinase-associated lipocalin levels predict mortality in elderly patients with chronic heart failure. Rejuvenation Res 2009;12:7–14.
59 Shrestha K, Borowski AG, Troughton RW, Thomas JD, Klein AL, Tang WH: Renal dysfunction is a stronger determinant of systemic neutrophil gelatinase-associated lipocalin levels than myocardial dysfunction in systolic heart failure. J Card Fail 2011;17: 472–478.
60 Jungbauer CG, Birner C, Jung B, Buchner S, Lubnow M, von Bary C, Endemann D, Banas B, Mack M, Boger CA, et al: Kidney injury molecule-1 and N-acetyl-β-D-glucosaminidase in chronic heart failure: possible biomarkers of cardiorenal syndrome. Eur J Heart Fail 2011;13:1104–1110.
61 Pitt B, Filippatos G, Gheorghiade M, Kober L, Krum H, Ponikowski P, Nowack C, Kolkhof P, Kim SY, Zannad F: Rationale and design of ARTS: a randomized, double-blind study of BAY 94-8862 in patients with chronic heart failure and mild or moderate chronic kidney disease. Eur J Heart Fail 2012; 14:668–675.
62 Acute Dialysis Quality Initiative (ADQI). www.ADQI.org (accessed January 10, 2013).

Dinna N. Cruz, MD, MPH
Division of Nephrology-Hypertension
University of California, San Diego, 200 West Arbor Drive 8409
San Diego, CA 92103-8409 (USA)
E-Mail dinnacruzmd@yahoo.com

McCullough PA, Kellum JA, Mehta RL, Murray PT, Ronco C (eds): ADQI Consensus on AKI Biomarkers and Cardiorenal Syndromes. Contrib Nephrol. Basel, Karger, 2013, vol 182, pp 137–157
DOI: 10.1159/000349971

Cardiorenal Syndrome Type 3: Pathophysiologic and Epidemiologic Considerations

Sean M. Bagshaw[a] · Eric A. Hoste[b] · Branko Braam[c] · Carlo Briguori[d] · John A. Kellum[e] · Peter A. McCullough[f] · Claudio Ronco[g] · for the Acute Dialysis Quality Initiative 11 Working Group

[a]Division of Critical Care Medicine, Faculty of Medicine and Dentistry, University of Alberta, Edmonton, Alta., Canada; [b]Department of Critical Care Medicine, University of Gent, Gent, Belgium; [c]Division of Nephrology, Departments of Medicine and Physiology, Faculty of Medicine and Dentistry, University of Alberta, Edmonton, Alta., Canada; [d]Department of Cardiology, Clinica Mediterranea, Naples, Italy; [e]Clinical Research, Investigation, and Systems Modeling of Acute Illness Center, Department of Critical Care Medicine, University of Pittsburgh, Pittsburgh, Pa., USA; [f]St. John Providence Health System, Warren, Mich., Providence Hospitals and Medical Centers, Southfield and Novi, Mich., St. John Macomb Oakland Center, Madison Heights, Mich., St. John Hospital and Medical Center, Detroit, Mich., USA; [g]Department of Nephrology Dialysis and Transplantation and International Renal Research Institute (IRRIV), San Bortolo Hospital, Vicenza, Italy

Abstract

Cardiorenal syndrome (CRS) type 3 is a subclassification of the CRS whereby an episode of acute kidney injury (AKI) precipitates and contributes to the development of acute cardiac injury. There is limited understanding of the pathophysiologic mechanisms of how AKI contributes to acute cardiac injury and/or dysfunction. An episode of AKI may have effects that depend on the severity and duration of AKI and that both directly and indirectly predispose to an acute cardiac event. Moreover, baseline susceptibility will modify the subsequent risk for cardiac events associated with AKI. Experimental data suggest cardiac injury may be directly induced by inflammatory mediators, oxidative stress, apoptosis and activation of neuroendocrine systems early after AKI. Likewise, AKI may be associated with physiologic derangements (i.e. volume overload; metabolic acidosis, retention of uremic toxins, hyperkalemia; hypocalcemia), alterations to coronary vasoreactivity, and ventricular remodeling and fibrosis that indirectly exert negative effects on cardiac function. AKI may also adversely impact cardiac function by contributing to alternations in drug pharmacokinetics and pharmacodynamics. Additional experimental and transla-

ADQI 11 Workshop members are listed in Appendix 2.

tional investigations coupled with epidemiologic surveys are needed to better explore that pathophysiologic mechanisms underpinning acute cardiac events associated with AKI and their impact on outcomes.

Cardiorenal syndrome (CRS) type 3 (acute renocardiac syndrome) is a subclassification of the CRS whereby an episode of acute kidney injury (AKI) precipitates and contributes to the development of acute cardiac injury and/or dysfunction [1]. There are a number of potential contributing causes for AKI that may predispose to the development of CRS type 3 and which are relevant for the susceptibility, etiology, severity and duration of the type 3 CRS (table 1).

In general, there is limited understanding of the pathophysiologic mechanisms whereby AKI contributes to acute cardiac injury and/or dysfunction. An episode of AKI may have severity and duration-dependent effects that both directly and indirectly predispose to an acute cardiac event. There is some experimental data to suggest that cardiac injury may be directly induced by inflammatory mediators, oxidative stress, and upregulation of neuroendocrine systems early after AKI [2–5]. Likewise, AKI may be associated with physiologic derangements (i.e. volume overload; metabolic acidosis, retention of uremic toxins, hyperkalemia; hypocalcemia), alterations to coronary vasoreactivity, and ventricular remodeling and fibrosis that indirectly exert negative effects on cardiac function [6–10]. The development of AKI may also adversely impact cardiac function, through both direct and indirect mechanisms, by contributing to alternations in drug pharmacokinetics and pharmacodynamics. The relationship between these direct and indirect mechanisms remains poorly understood.

There is a paucity of data describing the epidemiology of acute cardiac events (i.e. arrhythmia, myocardial ischemia, heart failure, cardiac arrest) occurring after an episode of AKI. There is also a lack of data specifically describing the temporal sequence of cardiac events after onset of AKI, including the dose-response effects related to AKI severity and duration on short- and long-term cardiac function [11, 12]. Moreover, there are challenges when trying to extrapolate on the epidemiology of CRS type 3 from the available literature due to: (1) variability in the methods for defining AKI and heterogeneity in the contributing causes for AKI; (2) variability in the populations at risk being studied and their susceptibility for development of either AKI or acute cardiac events; (3) difficulty in clarifying the temporal association between AKI and acute cardiac events, in particular from observational data, and (4) failure of many observational studies and clinical trials focused on AKI to specifically capture and/or describe the occurrence of acute cardiac events as key outcomes of interest.

Accordingly, the objectives of this review were to: (1) to design a pathophysiologic model of CRS type 3 supported by evidence; (2) identify from this

Table 1. Summary of potential contributing causes for AKI contributing to CRS type 3

Prototypical condition	
Contrast-induced AKI	Post-inflammatory GN
Drug-induced AKI	Rhabdomyolysis
Major surgery	Acute pyelonephritis
Cardiac surgery	Post-obstructive uropathy

Table 2. Summary of susceptibilities for CRS type 3

Risk-modifying factors	
Age	Congestive heart failure
Sex	Pulmonary disease
Coronary artery disease	Chronic kidney disease
Hypertension	Systemic vascular disease
Hypercholesterolemia	Systemic immune disease
Diabetes mellitus	Infection/sepsis

model descriptive/epidemiological and mechanistic knowledge gaps, and (3) provide where possible recommendations for future investigation.

Methods

The 11th Scientific Acute Dialysis Quality Initiative (ADQI) Meeting on Cardio-Renal Syndromes assembled experts on this topic, including cardiologists, nephrologists, intensivists, physiologists and epidemiologists [13].

This report is the result of a modified Delphi analysis performed by the ADQI Working Group [14]. The Delphi method is a structured and standardized process for collecting, summarizing and disseminating knowledge from a group of experts focused on a specific problem or task. A detailed description of the ADQI methodology is available at www.adqi.net [15]. In brief, the process included pre-meeting, at meeting and post-meeting activities.

Prior to the meeting, each member of the working subgroup on the topic of CRS type 3 developed a list of preliminary questions and objectives. The key questions were focused on two broad themes: understanding the pathophysiologic mechanisms and the epidemiology of CRS type 3 (table 2).

Each working group member subsequently performed a systematic literature search focused around these key questions. The literature search was conducted using the Medline database (via the PubMed interface). In general, three broad search themes were used: (1) 'pathophysiology'; (2) 'acute kidney injury' and 'acute renal failure', and (3) 'cardiorenal', 'myocardial infarction', 'heart failure', 'arrhythmia', and 'acute coronary syndrome'. The bibliographies of identified studies and relevant review articles or editorials were also reviewed along with the personal records of each participating member for additional potentially relevant studies. This literature was reviewed and evaluated to determine the current status of consensus on each of the key questions.

Activities at the meeting, involving the entire working group, included a review of the ADQI methodology, a systematic evaluation of the quality of the retrieved evidence, and incorporation of evidence-based medicine principles into the review process. If evidence on specific key questions was unavailable or lacking, the working subgroup discussed how 'best clinical practice' would be established. The working subgroup used breakout sessions from the main meeting to review, discuss and catalogue the available literature and formulate summary statements to describe existing consensus, best practice and recommendations for future research. The findings from these working subgroup breakout sessions were then presented to the entire ADQI delegate working group at the meeting for feedback and suggested revisions. This method was repeated through an iterative process until majority consensus was achieved.

Post-meeting activities included compilation of the findings and consensus from the meeting and drafting of summary documents.

Findings

Acute Kidney Injury

AKI is common and increasingly encountered in hospitalized patients. The majority of episodes of severe AKI occur in association with critical illness and are supported in an intensive care setting. Susceptibility to development to AKI likely stems from numerous interacting factors including genetic predisposition [16–19], demographics and burden of comorbid illness such as older age [20], cardiovascular disease, diabetes mellitus and chronic kidney disease (CKD), coupled with clinical events and/or iatrogenic exposures. In addition, the severity and duration of AKI correlated with subsequent risk for complications [21, 22]. However, few data have described the epidemiology of acute cardiac events (i.e. arrhythmia, myocardial ischemia, heart failure, cardiac arrest) temporally following an episode of AKI, including whether dose-response effects related to AKI severity and duration impact short- and long-term risk of cardiac events.

Pathophysiologic Mechanisms of AKI-Induced Cardiac Injury

Direct Mechanisms

AKI is accompanied by almost instantaneous non-hemodynamic events that originate in the kidney and have distant organ effects, including the heart [2, 23]. The direct physiologic mechanisms by which AKI interacts with the heart have been referred to as the 'cardiorenal connectors' [24]. These mechanisms encompass, but are not likely limited to activation of the systemic immune system (i.e. pro- and anti-inflammatory cytokines and chemokines with alteration in leukocyte trafficking and extravasation); the sympathetic nervous (SNS), renin-angiotensin-aldosterone (RAAS) and coagulation systems; along with increases in oxidative stress (i.e. disruption in the balance between radical oxygen species and nitric oxide (NO)). For each of these 'cardiorenal connector' mechanisms, there is limited experimental evidence to support a pathophysiologic role of AKI directly mediating cardiac events and/or dysfunction.

The majority of experimental studies evaluating the immune 'cardiorenal connectors' have focused on renal ischemia and ischemia/reperfusion injury (IRI). Accordingly, there is little description of the direct interaction of immune response and cardiac performance following AKI in alternative settings (i.e. nephrotoxin, post-obstructive, sepsis). In IRI models, AKI has been shown to elicit a systemic immune response characterized by dose-response increase in circulating levels of pro- and anti-inflammatory mediators. AKI may also contribute to functional alterations in immune cell responsiveness and alterations to leukocyte trafficking, adhesion and tissue extravasation both locally in the kidney and in distal organs such as the heart [2]. Leukocytes have long been thought important in cardiac dysfunction following ischemia and blocking leukocyte function or localization may protect against injury [25]. Elevated white blood cell count is associated with increased risk of acute myocardial infarction [26] and removing white blood cells from cardioplegia solution improves myocardial function in stressed piglet hearts [27].

Several studies have found circulating levels of tumor necrosis factor-α (TNF-α), interleukin (IL)-1 and IL-6 to increase immediately after renal IRI in experimental models [3–5, 7, 28, 29]. Several cytokines have now been implicated as myocardial depressant substances including TNF-α, IL-1β, IL-6, IL-2, and interferon-γ (IFN-α). Elevated TNF-α and IL-6 have been shown to have direct cardiodepressant effects, characterized by reduced left ventricular (LV) fractional shortening, and elevated LV end-diastolic and systolic dimensions [2, 3, 24, 30]. Cytokines can impact myocardial function due to both impaired myocyte contractility and interaction with the extracellular matrix. In general, cytokine-mediated contractile dysfunction is reversible over an extended time period of several days following exposure [4]. Although these cytokines typically impart negative inotropic effects, the nature and pattern of the inotropic response is more complex, consisting of an immediate response (within minutes) that can be either stimulatory or depressant, depending on the experimental conditions and physiologic milieu, and a delayed response lasting hours to days that is uniformly cardiodepressant and dependent on the production of secondary mediators [4]. The cellular mechanisms underlying these effects are multiple. NO derived from constitutive NO synthase, sphingolipid mediators, arachidonic acid, and alterations in intracellular Ca^{2+} handling have all been shown to contribute to the immediate contractile effects, whereas the delayed response results primarily from the combined influence of NO generated from inducible NOS, the production of radical oxygen species and alterations in β-adrenergic receptor signaling [4].

In the isolated rat heart, perfusion with TNF-α results in a decrease in LV diastolic pressure and coronary vasoconstriction [31]. Sustained infusion of TNF-α causes time-dependent, delayed (2 h to 2 days) contractile dysfunction and LV dilatation, that could persist for up to 10 days, depending on the dose administered [5]. Reduced LV performance has also been associated with several diastolic abnormalities including slowing of relaxation, LV dilatation without changes in end-diastolic pressure (myocardial creep), and a leftward shift of the end-diastolic pressure-strain relation, indicating increased diastolic stiffness. Pressure-volume studies of the isolated, blood-perfused canine heart have shown that TNF-α acutely impacts LV mechanoenergetics by increasing the oxygen cost of contractility, an effect ascribed to alterations in energy utilization for excitation-contraction coupling [32].

Following IRI, rat hearts also exhibit increased expression of intercellular adhesion molecule-1, myeloperoxidase and apoptosis [3]. Of interest, only renal IRI and not bilateral nephrectomy resulted in a significant increase in myocardial apoptosis, implying systemic inflammation rather than acute uremia may play a more immediate and direct role in myocardial injury and dysfunction in AKI. Moreover, TNF-α likely contributes directly to cardiac myocyte apoptosis, as demonstrated by attenuation of apoptosis following administration of anti-TNF-α antibodies [3].

Increased circulating levels of pro-inflammatory cytokines have been demonstrated in patients with chronic heart failure and are associated with maladaptive ventricular remodelling, poor functional status, chronic cachexia and death [33]. Two large randomized trials have evaluated of anti-TNF therapies for at-risk chronic heart failure patients, however while shown to be relatively safe, neither were efficacious for reducing mortality [34]. To date, TNF-α inhibition following an episode of AKI and its effects are myocardial injury and/or function have not yet been evaluated. By extrapolation, this concept may have merit for CRS type 3, in settings where moderate to severe AKI induces a strong immune response characterized by markedly elevated levels of inflammatory mediators, in particular those with direct myocardial effects such as TNF-α. As an example, in a folic-acid-induced rat model of AKI, administration of a single low dose of cyclosporine A (1–5 mg/kg) not only markedly attenuated development of AKI (i.e. rise in serum creatinine, urea) but also attenuated systemic IL-6 expression, TNF-like weak inducer of apoptosis (TWEAK) expression, nuclear factor-κB activation, inflammatory cell infiltration and renal tubular cell apoptosis when compared with controls [35]. Recently, deficiency of heme oxygenase-1, an enzyme induced by oxidative stress and responsible for degradation of heme molecules, was found to exacerbate induction of systemic IL-6 along with IL-6 mRNA and downstream signaling effectors in myocardial tissue [28].

Alternatively, milder forms of renal IRI may have a 'protective' effect on the heart against ischemic injury. In a rat study of brief (15 min) renal artery occlusion under hypothermic conditions (30 °C), Gho et al. [36] found marked reductions in myocardium at risk and total area infarction when compared with controls, however renal IRI injury was failed to attenuate injury at normothermia. The precise mechanisms by which mild renal IRI may protect against downstream myocardial ischemia are incompletely understood, however may in part be mediated by alterations in sympathetic outflow coupled with upregulation of anti-inflammatory mediators and anti-apoptotic responses [36, 37].

There is limited data available describing the role of neuroendocrine activation, specifically the SNS and RAAS in the pathophysiologic of CRS type 3. However, activation of the SNS is a hallmark of both AKI and acute HF [38]. The enhancement of renal SNS activity and its consequent effect on norepinephrine overflow from nerve terminals during AKI [39] may impair myocardial function through several mechanisms, including: (1) direct effects of norepinephrine, (2) disturbances in myocardial Ca^{2+} homeostasis, (3) increase in myocardial oxygen demand contributing to susceptibility to endocardial ischemia, (4) cardiac myocyte apoptosis mediated through β_1-adrenergic receptors stimulation [40], (5) stimulation of α_1-adrenergic receptors, contributing to cardiac myocyte hypertrophy, and (6) direct activation of the RAAS. Heightened adrenergic drive can stimulate β_1-adrenergic receptors in the juxtaglomerular apparatus of the kidneys contributing to reduced renal blood flow and exacerbated release of renin and RAAS activation. Maladaptive RAAS activation in AKI contributes to angiotensin II (ANGII) release, vasoconstriction and further loss of extracellular fluid homeostasis. ANGII may redundantly contribute to systemic vasoconstrictor and excessive elevation of the systemic vascular resistance. ANGII may also play a direct role in modifying myocardial structure and function [41], contribute to cellular hypertrophy, and precipitate apoptosis in cardiac myocyte cultures [42]. ANGII is also a potent stimulator of a number of cell signaling pathways including those involved in oxidative stress, inflammation, and the regulation of the extracellular matrix [43].

In a dog model of renal IRI, Kingma et al. [8] implied increased activity of the RAAS contributing to observed diminished coronary responsiveness to acetylcholine, adenosine, bradykinin, and L-arginine. In addition, renal IRI resulted in greater myocardial oxygen consumption at rest. While a definitive role for the RAAS was not established, these data imply that AKI may directly contribute to impaired coronary vasoreactivity and elevated myocardial oxygen consumption that would potentially increase susceptibility to myocardial ischemia and major cardiac events. Several rat models of renal IRI have shown increased secondary lung injury due largely to alternation in pulmonary vascular permeability with resultant interstitial edema and alveolar hemorrhage [2]. The altered pulmonary vascular permeability associated with AKI is directly mediated by systemic inflammatory mediators and kidney-lung cross-talk. The resulting alveolar congestion and edema is in part mediated by altered expression of epithelial sodium channel and aquiporin-5 expression. Similar mechanisms may also contribute to myocardial interstitial edema and dysfunction [44]. Transgenic sickle mice subjected to renal IRI exhibited significantly increased myocardial capillary congestion, serum amyloid-P (murine equivalent of C-reaction protein), and renal expression of caspase-3 compared with controls [45].

In addition, while we considered the cardiac-specific cellular responses associated with AKI including the role of mitochondrial dysfunction, apoptosis, remodelling, and fibrosis, limited data were available. Robinson et al. [46] evaluated cardiac mitochondrial function in a rat model of glycerol-induced AKI. In this study, isolated heart preparations showed site-unspecific uncoupling of oxidative phosphorylation. This contributed to increased mitochondrial Ca^{2+} uptake, and may provide explanation for the diminished inotropic response in the setting of AKI.

Cardiac myocyte apoptosis and neutrophil infiltration are important contributors to the pathophysiology of myocardial infarction during AKI and transgenic models have shown that even apoptosis alone can lead to lethal heart failure [3]. In this study, kidney IRI, but not uremia, was critical to induction of apoptosis in myocardial tissue. Another experimental model of cisplatin-induced AKI found an increased number of TUNEL-positive cells implying significant myocardial apoptosis [47]. It has been demonstrated that apoptotic rates of only 23 cardiac myocytes per 10^5 nuclei are sufficient to induce a lethal, dilated cardiomyopathy [48]. Although this frequency of cell death is quite low, it is still 15-fold higher than that observed in healthy human hearts. Decreasing intracellular iron levels is also known to induce apoptosis [49].

The strongest evidence of a cardiorenal link between AKI and the development of cardiac fibrosis is the β-galactoside-binding lectin galectin-3 [9, 10]. Galectin-3 mRNA expression in renal tubules was shown to be upregulated early after IRI and toxin-induced AKI and persisted for 7 days following injury [9]. Importantly, galectin-3 has been implicated in the development of myocardial fibrosis and heart failure in an experimental model of AKI [10]. Moreover, inhibition of galectin-3 can mitigate the formation of cardiac fibrosis [50].

Indirect Mechanisms

The workgroup identified a number of potential consequences of AKI that may act to indirectly contribute to reduced cardiac function (fig. 1, 2).

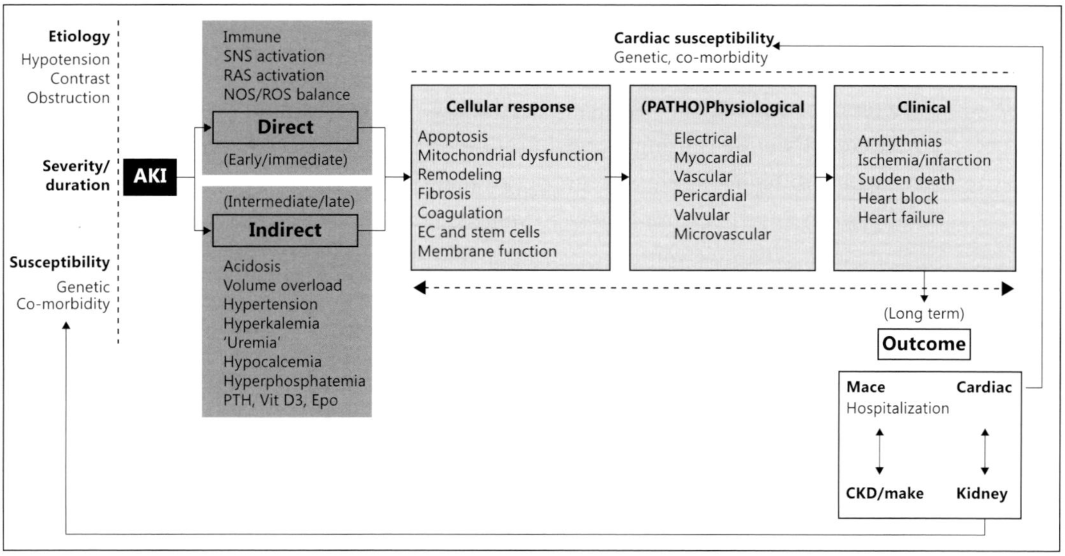

Fig. 1. Summary of the demographic contributors, clinical susceptibilities and pathophysiologic mechanisms for development of CRS type 3. Reproduced with permission from ADQI [88].

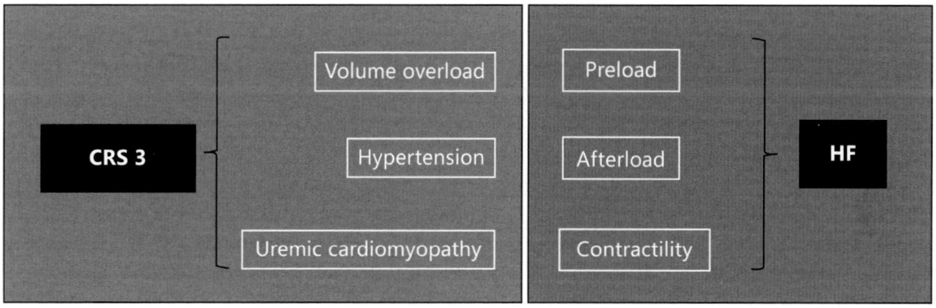

Fig. 2. Summary of the interaction of clinical complications of AKI and theoretical negative physiologic effects on myocardial performance. Reproduced with permission from ADQI [88].

Volume Overload. Excess fluid accumulation is associated with less favorable outcomes in critically ill patients. In a large randomized trial of fluid management strategies in critically ill patients with acute lung injury, a conservative fluid regimen was associated with improved lung function and a trend for lower utilization of renal replacement therapy (RRT) compared with a liberal fluid regimen [51]. Similar data have also been shown in critically ill children [52]. Moreover, excess fluid accumulation may worsen kidney function through venous congestion and intra-abdominal hypertension. Several observational studies, while not established causality, consistently show association between fluid overload and less favorable outcomes in critically ill patients with AKI [53, 54]. In addition, fluid removal in these patients, either through loop diuretics use [55] or ultrafiltration [56], was associated with higher survival. Fluid accumulation may adversely impact myocardial performance [44], predispose to ventricular arrhythmias [57] and may contribute to maladaptive ventricular remodelling and myocardial fibrosis [44, 58].

Hypertension. The majority of critically ill patients with AKI may present with hypotension due to the precipitating etiology for AKI (i.e. major surgery, sepsis, cardiogenic shock), however beyond the inciting event contributing to CRS type 3, a number of mechanisms may contribute to the development of both systemic and pulmonary hypertension. This may occur through volume overload contributing to elevated preload, alterations in cardiac output and increased afterload through total body autoregulation, activation of the SNS and RAAS. In susceptible patients, these may contribute to myocardial ischemia, arrhythmias and ventricular dysfunction. Increased activation of the SNS and RAAS has been well described in conditions such as hepatorenal syndrome and acute decompensated heart failure [59].

Acidemia. Metabolic acidosis is a common complication of AKI. In a recent survey of critical care clinicians, 74% of the respondents stated that metabolic acidosis was a key factor contributing to a decision to initiate RRT among critically ill patients [60]. In the RENAL study, a large prospective randomized study evaluating the intensity of continuous RRT (CRRT), 35% of critically ill patients had severe acidemia at time of initiation of CRRT [61]. The accumulated acid (i.e. H^+ ions) may alter the protein structure and interfere with normal function. This may directly contribute to decreased myocardial contractility through changes to β-receptor expression and altered intracellular calcium handling [62].

Hyperkalemia. Similar to acid-base homeostasis, potassium is predominantly dependent on kidney function for elimination from the body. Hyperkalemia is a life-threatening complication of AKI and an important absolute indication for RRT [60], however studies of patients with severe AKI show hyperkalemia is a relatively infrequent reason for RRT initiation [63]. Hyperkalemia is associated with cardiac toxicity and arrhythmic complications (see below).

Hypophosphatemia. While untreated AKI is often associated with elevated plasma phosphate levels, patients started on RRT commonly develop hypophosphatemia [61, 64, 65]. Hypophosphatemia may be an unrecognized iatrogenic complication associated with respiratory muscle weakness and impaired myocardial performance [66, 67].

Uremic Toxins. In CKD, more than 100 uremic toxins or retention products have been identified [68]. Uremic toxins include water-soluble, protein-bound and small molecular weight compounds with variable pharmacodynamic profiles. Several uremic toxins (i.e. indoxylsulfate, *p*-cresol conjugates, β_2-microgliobulin, and FGF-23) are associated with cardiovascular toxicity in CKD [69]. However, there is a relative paucity of data on acute uremia in AKI. In AKI, uremic compounds may exert different time-dependent physiologic effects when compared to those described in CKD. There is indirect evidence to suggest uremic compounds in AKI may negatively impact kidney function. For example, coronary vasoreactivity is adversely affected in AKI, resulting in altered thresholds for myocardial ischemia in conditions of increased myocardial oxygen demand [8]. Uremic toxicity is complicated, as varying duration and severity of AKI as well as concomitant non-renal organ dysfunctions may synergistically act to adversely impact cardiac function.

Pathophysiological Effects of AKI on the Heart

The pathophysiologic mechanisms underlying the interplay between kidney injury and heart dysfunction are still not completely understood (fig. 2). However, we can summarize several of the physiologic effects whereby AKI may negatively impact heart function as: (a) electrophysiological, (b) ischemic, (c) myocardial and/or (d) pericardial.

Electrophysiological Effects

The loss of acid-base, electrolyte and fluid homeostasis occurring in AKI may precipitate alteration in heart rhythm. Electrical signaling in the heart involves the passage of ions through ionic channels. The Na^+, K^+, Ca^{2+}, and Cl^- ions are the major charge carriers, and their movement across the cell membrane creates a flow of current that generates excitation and signals contraction in cardiac myocytes.

Hyperkalemia is associated with a distinctive sequence of electrocardiogram (ECG) changes. The earliest is usually narrowing and peaking (tenting) of the T wave (fig. 3). The ratio of intracellular to extracellular potassium is important for determining the cellular membrane potential. Small changes in the extracellular potassium level can have profound effects on the function of the electrical conduction system of the heart. In the setting of hyperkalemia, the excess extracellular potassium will increase the activity of the potassium channel, resulting in faster repolarization and thus creating the narrow base and tall T wave. On the other hand, this will inactivate the sodium channel, preventing proper conduction of electrical activity. Progressive extracellular hyperkalemia reduces atrial and ventricular resting membrane potentials, thereby inactivating sodium channels. Moderate to severe hyperkalemia occasionally induces ST segment elevations in the right precordial leads (V1 and V2) and simulates an ischemic

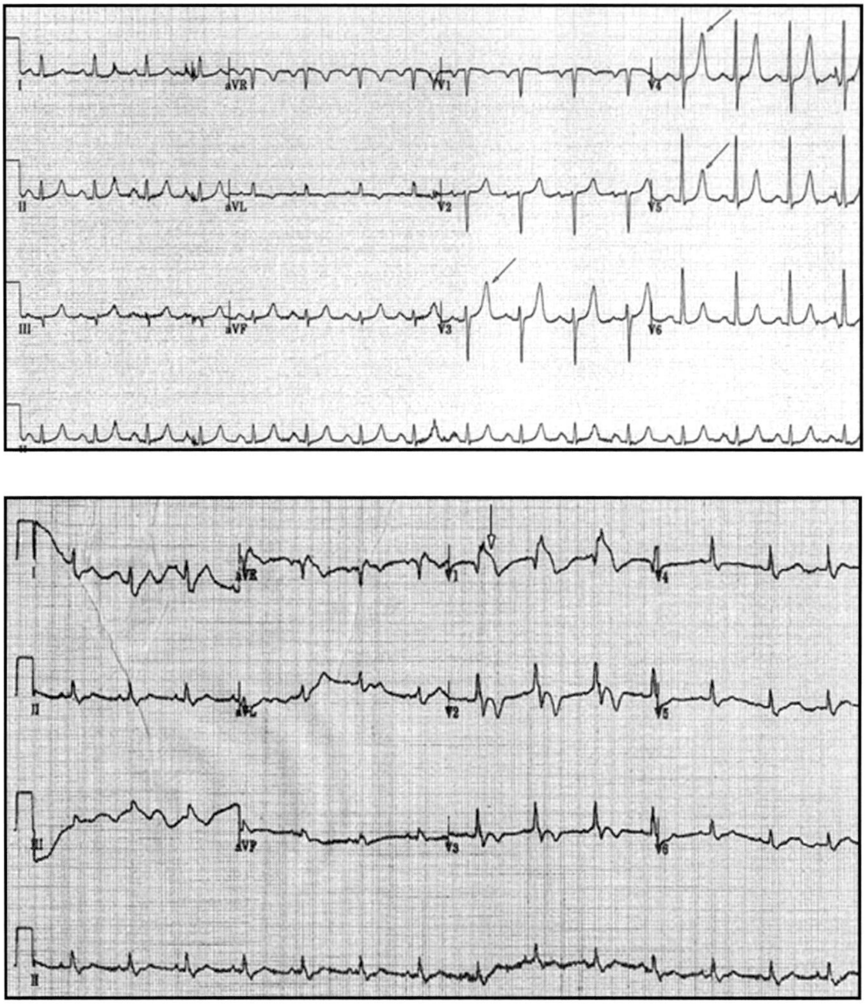

Fig. 3. Hyperkalemia. **a** A 54-year-old woman with acute renal failure due to glomerulonephritis status was admitted with acute onset of shortness of breath. Her clinical examination and radiographic findings were consistent with acute pulmonary edema. ECG shows normal sinus rhythm with tall T waves best seen in leads V3, V4 and V5. The tall T wave is symmetrical, narrow, scooped inwards and 'tented'. K^+ level was 5.8 mEq/l. This classic appearance of the T wave is usually associated with hyperkalemia. The PR interval, QRS duration and QTc interval are all within normal range. The P wave is wide (>0.12 s in duration) and bifid in shape – so-called 'P-mitrale' in lead II. There is a negative component in lead V1, which is widened and slurred, consistent with left atrial enlargement. **b** ECG shows significant widening of the QRS complex to 158 ms with right bundle branch block morphology associated with coved ST segment elevation in lead V1, resembling 'Brugada ECG pattern', which can be seen in severe hyperkalemia. Reproduced with permission from ADQI [88].

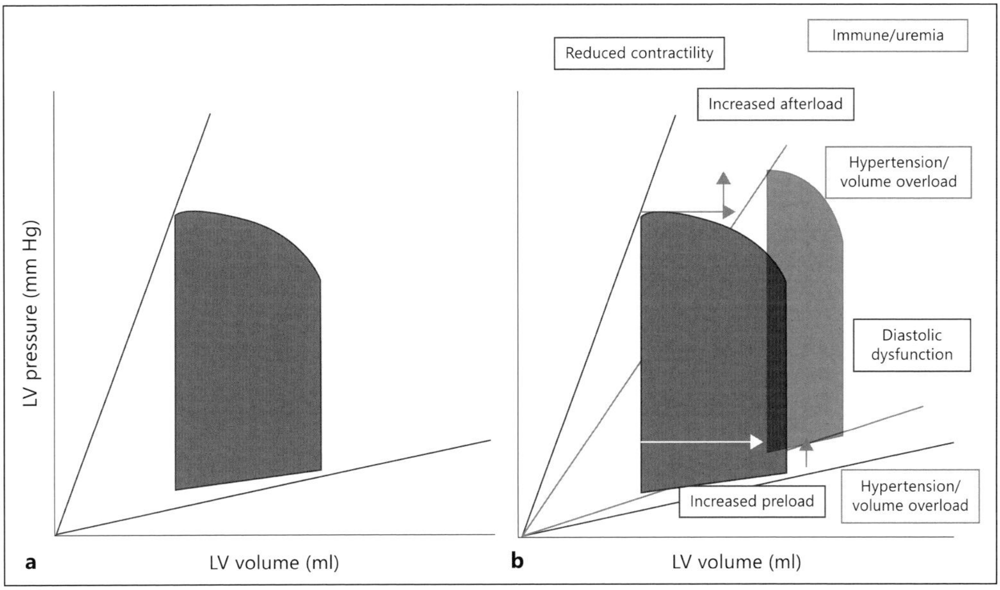

Fig. 4. a, b Mechanisms of LV dysfunction in CRS type 3. Reproduced with permission from ADQI [88].

current-to-injury pattern. However, even severe hyperkalemia can be associated with atypical or non-diagnostic ECG findings. Hyperkalemia can induce a Brugada-like pattern in the ECG [70]. This usually occurs in critically ill patient with significant hyperkalemia (serum [K^+] >7.0 mmol/l) and is associated with pseudo-right bundle branch block and persistent ST segment elevation in at least two precordial leads (fig. 4). Very marked hyperkalemia leads to eventual asystole, sometimes preceded by a slow undulatory (sine-wave) ventricular flutter-like pattern [71].

Severe hypermagnesemia can cause AV nodal and intraventricular conduction disturbances that may culminate in complete heart block and cardiac arrest.

An increased extracellular calcium concentration shortens the ventricular action potential duration by shortening phase 2 of the action potential. This cellular change correlates with the shortening of the QT interval (ST segment portion). Hypercalcemia may produce a high takeoff of the ST segment in leads V1 and V2 that can simulate acute ischemia. Ca^{2+} plays a central role in the regulation of myocardial contraction and relaxation and there is increasing evidence that disturbances in Ca^{2+} handling may play a central role in the disturbed contractile function in heart failure. The intracellular Ca^{2+} transient demonstrates a blunted rise with depolarization, reflecting slower delivery of Ca^{2+} to the contractile apparatus (causing slower activation) and a slower rate of fall during repolarization (causing slower relaxation). These two abnormalities could explain both systolic and diastolic dysfunction [72].

Ischemic Effects

The theoretical pathophysiological mechanisms responsible of myocardial ischemia during AKI may also include the following: (1) acidemia; (2) neurohormonal activation (i.e. SNS and RAAS), and (3) acute accumulation of uremic toxins and cytokines [31]. The SNS activation can lead to vasoconstriction and increased myocardial oxygen demand. These can lead to ischemia when oxygen delivery is restricted, in particular in susceptible patients. TNF-α induces coronary vasoconstriction, which contributes indirectly to contractile dysfunction in isolated, crystalloid-perfused rat hearts [73]. Endothelin and bradykinin may be also responsible for the reduced vessel function. Endothelin is a potent peptide vasoconstrictor released by endothelial cells throughout the circulation. Its release from endothelial cells can be enhanced by several vasoactive agents (ANGII, norepinephrine) and cytokines (TNF, and IL-1β). During AKI, higher levels of myocardial oxygen demand would trigger maldistribution of ventricular blood flow and result in subendocardial ischemia. It has been demonstrated that during AKI (1) coronary autoregulation is preserved but the lower autoregulatory break point shifts to a higher coronary perfusion pressure; (2) coronary vascular reserve and coronary vascular conductance are markedly diminished, and (3) coronary vessel reactivity to either endothelium-dependent or -independent vasodilators is substantially reduced [8]. The observed reduction in coronary venous PO_2 during AKI may reflect: (1) a reduced bioavailability of NO to vascular smooth muscle cells as a result of structural alterations in the vessel or microvessel wall, (2) an increased destruction of NO as a result of overproduction of reactive oxygen intermediates in the vessel wall, and (3) an impaired dilatation to NO donors (e.g. nitroglycerin) as a result of defective vascular smooth muscle cell function at the level of the soluble guanylate cyclase/cyclic guanosine monophosphate signaling pathway [74].

Recent evidence has emerged to implicate a potentially important pathophysiologic link between neutrophil gelatinase-associated lipocalin (NGAL), a 25-kDa lipocalin covalently bound to gelatinase from human neutrophils involved inflammation and cellular matrix degradation, and CRSs [75]. NGAL is a strong marker of acute renal tubular injury [76]. Renal NGAL expression rapidly increased following acute inflammation and/or injured renal tubular epithelia, in particular following damage from IRI and toxin exposure [77]. NGAL is also believed to play an important role in iron homeostasis and traffic and may activate apoptotic pathways by regulating the intracellular iron concentration

[78]. NGAL has also been implicated in the pathophysiology of heart failure, myocarditis and in coronary atherosclerosis [79–81]. Likewise, urinary kidney injury molecule-1, a specific marker of renal tubular damage, has also been implicated for modifying the susceptibility to heart failure risk [82]. These data suggest that acutely upregulated biomarkers of kidney damage, such as NGAL and kidney injury molecule-1, may circulate to have distant pathophysiologic effects on the heart, including myocardial inflammation, destabilization of atherosclerotic plaques and susceptibility to ischemic events.

Myocardial Effects

In animal experiments, it has been shown that 48 h after kidney IRI, functional changes in the myocardium including LV dilatation, increased LV end-diastolic and end-systolic diameters, increased relaxation time, and decreased fractional shortening are evident [3]. The proposed pathophysiological mechanisms responsible for LV dysfunction following an episode of AKI include accumulation of inflammatory mediators (i.e. cytokines); leukocyte infiltration, SNS activation; acute sodium and fluid retention; metabolic acidosis, and accumulation of uremic toxins.

The ventricular pressure-volume loop provides a paradigm to understand the relationship between intraventricular pressure and volume throughout the cardiac cycle. Figure 4 shows the pressure-volume relationship under normal conditions and following development of severe AKI. The combined global effect was therefore postulated. The theoretical detrimental effects of AKI on myocardial mechanics may include: (1) increase in preload due to intravascular fluid accumulation (causes a rightward shift of the LV loop); (2) increase in afterload due to increased mean systemic blood pressure (causes an increase in LV end-systolic pressure); (3) increase in diastolic dysfunction due to loss of calcium homeostasis (causes an upward shift of the LV loop), and (4) increase in systolic dysfunction due to inflammatory mediators and uremic toxins (causes a reduction in contractility).

Pericardial Effects

Uremia through the myocardial-depressant factors (cytokines) may precipitate pericarditis [83].

Clinical Outcomes Associated with CRS Type 3

There is limited data in the literature on the impact of AKI on occurrence of cardiac-specific events and outcomes (table 3). However, in a large population-based cohort study, James et al. [84] evaluated the association of AKI following coronary angiography (i.e. CI-AKI) and occurrence of cardiovascular-related

Table 3. Summary of key questions proposed by the ADQI Working Group

Pathophysiologic mechanisms of CRS type 3 1. What are the most important 'direct' pathophysiologic mechanisms following an episode of AKI that contribute to cardiac injury/dysfunction? a. What is the time course (temporality)? b. What is the impact of etiology, severity and/or duration (dose response)?
2. What are the most important 'indirect' pathophysiologic mechanisms following an episode of AKI that contribute to cardiac injury/dysfunction? a. What is the time course (temporality)? b. What is the impact of etiology, severity and/or duration?
3. What is the relationship between direct and indirect pathophysiologic mechanisms on cardiac injury/dysfunction? a. What is the relationship between cardiac pressure-volume mechanics (i.e. indirect) and myocardial function (i.e. direct) following an episode of AKI?
Epidemiology of CRS type 3 1. What are the important factors modifying the risk/susceptibility for cardiac-specific events after an episode of AKI? a. Baseline susceptibilities kidney and cardiac-specific factors (non-modifiable)?
2. What are the key contributing factors for development of AKI in CRS type 3? a. What is an ideal representative etiology for AKI to describe the CRS type 3 (i.e. contrast-induced nephropathy) b. What is the role iatrogenesis in CRS type 3?
3. What is the incidence and spectrum of cardiac-specific events following an episode of AKI?
4. What is the time course for cardiac-specific events following an episode of AKI? (ESC guidelines ~1 month)
5. What are the implications of CRS type 3 on the natural history of cardiac function?

hospitalizations, in addition to mortality and end-stage kidney disease (ESKD). Increasing severity of CI-AKI was associated with a dose-response increase in risk of heart failure hospitalization (AKIN stage I: HR 1.48, 95% CI 1.16–1.91; AKIN stage II/III: 2.17, 95% CI 1.49–3.15). Similar patterns were also seen for mortality, progression to ESKD and AKI-related hospitalization. Anzai et al. [12] prospectively evaluated 141 consecutive patients receiving reperfusion therapy after presentation with first anterior ST-elevation myocardial infarction, of which 22.0% (n = 31) developed AKI within 48 h, defined as an increase in serum creatinine >0.3 mg/dl (26.5 μmol/l). The development of early post-reperfusion AKI was independently associated with higher risk for in-hospital cardiac events including heart failure, ventricular arrhythmia, and cardiac-specific death, along with long-term cardiac events including rehospitalization for heart failure and major adverse cardiac events (MACE). Cardiovascular-related deaths are the second most common cause of death after sepsis, predominantly attributable to heart failure and ischemic cardiac events, in patients with AKI [85].

Table 4. Summary of cardiac-specific events

Spectrum of outcome events
Heart failure
Myocardial infarction
Ventricular arrhythmia
Major adverse cardiac event (MACE)
Accelerated loss cardiac function
Cardiac surgery
Mechanical support device
Death
Hospitalization

Consensus Recommendations for Future Research

Future investigations focused on CRS type 3 should surveillance for cardiac-specific events (table 4) not only during the in-hospital period but also during long-term follow-up (i.e. 6–12 months). CRS type 3 may be associated with both downstream cardiac and kidney-specific adverse outcomes. In addition to cardiac-specific events, the occurrence of kidney-specific outcomes, including subsequent episodes of AKI [86], incident CKD, progression to ESKD and MACE should be captured. Future investigations should also integrate an assessment of baseline susceptibility for cardiac-specific events following development of CRS type 3 (i.e. low, intermediate, high) and evaluate how these susceptibilities are modified by either the severity of AKI and the duration of AKI. For example, risk of a cardiac-specific event associated with AKI may be similar for a patient suffering a mild but extended episode of AKI compared with a patient developing a more severe episode of relative short duration.

Consensus Recommendations for Clinical Practice

Evolving data suggest that surrogate measures of kidney-related damage, such as hypertension and proteinuria, may represent simply screening measures to predict long-term decrements in kidney function [87].

Conclusions

CRS type 3 represents a poorly understood subgroup of heart-kidney interaction, both with respect to the precise pathophysiologic mechanisms and its epidemiology. Additional experimental and translational investigations coupled with epidemiologic surveys are needed to better explore that pathophysiologic mechanisms underpinning acute cardiac events associated with AKI and their impact on outcomes.

References

1 Ronco C, Haapio M, House AA, Anavekar N, Bellomo R: Cardiorenal syndrome. J Am Coll Cardiol 2008;52:1527–1539.
2 Yap SC, Lee HT: Acute kidney injury and extrarenal organ dysfunction: new concepts and experimental evidence. Anesthesiology 2012;116:1139–1148.
3 Kelly KJ: Distant effects of experimental renal ischemia/reperfusion injury. J Am Soc Nephrol 2003;14:1549–1558.
4 Prabhu SD: Cytokine-induced modulation of cardiac function. Circ Res 2004;95:1140–1153.
5 Bozkurt B, Kribbs SB, Clubb FJ Jr, et al: Pathophysiologically relevant concentrations of tumor necrosis factor-α promote progressive left ventricular dysfunction and remodeling in rats. Circulation 1998;97:1382–1391.
6 Zager RA: Uremia induces proximal tubular cytoresistance and heme oxygenase-1 expression in the absence of acute kidney injury. Am J Physiol Renal Physiol 2009;296:F362–F368.
7 Zager RA, Johnson AC, Lund S: Uremia impacts renal inflammatory cytokine gene expression in the setting of experimental acute kidney injury. Am J Physiol Renal Physiol 2009;297:F961–F970.
8 Kingma JG Jr, Vincent C, Rouleau JR, Kingma I: Influence of acute renal failure on coronary vasoregulation in dogs. J Am Soc Nephrol 2006;17:1316–1324.
9 Nishiyama J, Kobayashi S, Ishida A, et al: Up-regulation of galectin-3 in acute renal failure of the rat. Am J Pathol 2000;157:815–823.
10 Sharma UC, Pokharel S, van Brakel TJ, et al: Galectin-3 marks activated macrophages in failure-prone hypertrophied hearts and contributes to cardiac dysfunction. Circulation 2004;110:3121–3128.
11 Chuasuwan A, Kellum JA: Cardio-renal syndrome type 3: epidemiology, pathophysiology, and treatment. Semin Nephrol 2012;32:31–39.
12 Anzai A, Anzai T, Naito K, et al: Prognostic significance of acute kidney injury after reperfused ST-elevation myocardial infarction: synergistic acceleration of renal dysfunction and left ventricular remodeling. J Card Fail 2010;16:381–389.
13 Ronco C, McCullough P, Anker SD, et al: Cardio-renal syndromes: report from the consensus conference of the Acute Dialysis Quality Initiative. Eur Heart J 2010;31:703–711.
14 Ronco C, Kellum JA, Mehta R: Acute Dialysis Quality Initiative (ADQI). Nephrol Dial Transplant 2001;16:1555–1558.
15 Kellum JA, Bellomo R, Ronco C: Acute Dialysis Quality Initiative (ADQI): methodology. Int J Artif Organs 2008;31:90–93.
16 Perianayagam MC, Tighiouart H, Liangos O, et al: Polymorphisms in the myeloperoxidase gene locus are associated with acute kidney injury-related outcomes. Kidney Int 2012;82:909–919.
17 Payen D, Lukaszewicz AC, Legrand M, et al: A multicentre study of acute kidney injury in severe sepsis and septic shock: association with inflammatory phenotype and HLA genotype. PLoS One 2012;7:e35838.
18 Frank AJ, Sheu CC, Zhao Y, et al: BCL2 genetic variants are associated with acute kidney injury in septic shock. Crit Care Med 2012;40:2116–2123.
19 Lu JC, Coca SG, Patel UD, Cantley L, Parikh CR: Searching for genes that matter in acute kidney injury: a systematic review. Clin J Am Soc Nephrol 2009;4:1020–1031.
20 Coca SG, Cho KC, Hsu CY: Acute kidney injury in the elderly: predisposition to chronic kidney disease and vice versa. Nephron Clin Pract 2011;119(suppl 1):c19–c24.
21 Brown JR, Kramer RS, Coca SG, Parikh CR: Duration of acute kidney injury impacts long-term survival after cardiac surgery. Ann Thorac Surg 2010;90:1142–1148.
22 Chawla LS, Amdur RL, Amodeo S, Kimmel PL, Palant CE: The severity of acute kidney injury predicts progression to chronic kidney disease. Kidney Int 2011;79:1361–1369.
23 Grams ME, Rabb H: The distant organ effects of acute kidney injury. Kidney Int 2012;81:942–948.
24 Bongartz LG, Cramer MJ, Doevendans PA, Joles JA, Braam B: The severe cardiorenal syndrome: 'Guyton revisited'. Eur Heart J 2005;26:11–17.

25 Ma XL, Lefer DJ, Lefer AM, Rothlein R: Coronary endothelial and cardiac protective effects of a monoclonal antibody to intercellular adhesion molecule-1 in myocardial ischemia and reperfusion. Circulation 1992;86:937–946.
26 Ernst E, Hammerschmidt DE, Bagge U, Matrai A, Dormandy JA: Leukocytes and the risk of ischemic diseases. JAMA 1987;257: 2318–2324.
27 Kronon MT, Allen BS, Halldorsson A, Rahman S, Wang T, Ilbawi M: Dose dependency of L-arginine in neonatal myocardial protection: the nitric oxide paradox. J Thorac Cardiovasc Surg 1999;118: 655–664.
28 Tracz MJ, Juncos JP, Croatt AJ, et al: Deficiency of heme oxygenase-1 impairs renal hemodynamics and exaggerates systemic inflammatory responses to renal ischemia. Kidney Int 2007;72:1073–1080.
29 Hirota H, Yoshida K, Kishimoto T, Taga T: Continuous activation of gp130, a signal-transducing receptor component for interleukin-6-related cytokines, causes myocardial hypertrophy in mice. Proc Natl Acad Sci USA 1995;92:4862–4866.
30 Blake P, Hasegawa Y, Khosla MC, Fouad-Tarazi F, Sakura N, Paganini EP: Isolation of 'myocardial depressant factor(s)' from the ultrafiltrate of heart failure patients with acute renal failure. ASAIO J 1996;42:M911–M915.
31 Edmunds NJ, Lal H, Woodward B: Effects of tumour necrosis factor-α on left ventricular function in the rat isolated perfused heart: possible mechanisms for a decline in cardiac function. Br J Pharmacol 1999;126:189–196.
32 Miyano H, Shishido T, Kawada T, et al: Acute effect of tumor necrosis factor-α is minimal on mechanics but significant on energetics in blood-perfused canine left ventricles. Crit Care Med 1999;27:168–176.
33 Rauchhaus M, Doehner W, Francis DP, et al: Plasma cytokine parameters and mortality in patients with chronic heart failure. Circulation 2000;102:3060–3067.
34 Anker SD, Coats AJ: How to RECOVER from RENAISSANCE? The significance of the results of RECOVER, RENAISSANCE, RENEWAL and ATTACH. Int J Cardiol 2002;86:123–130.
35 Wen X, Peng Z, Li Y, et al: One dose of cyclosporine A is protective at initiation of folic acid-induced acute kidney injury in mice. Nephrol Dial Transplant 2012;27: 3100–3109.
36 Gho BC, Schoemaker RG, van den Doel MA, Duncker DJ, Verdouw PD: Myocardial protection by brief ischemia in noncardiac tissue. Circulation 1996;94:2193–2200.
37 Verdouw PD, Gho BC, Koning MM, Schoemaker RG, Duncker DJ: Cardioprotection by ischemic and nonischemic myocardial stress and ischemia in remote organs. Implications for the concept of ischemic preconditioning. Ann NY Acad Sci 1996;793:27–42.
38 Kobuchi S, Tanaka R, Shintani T, et al: Mechanisms underlying the renoprotective effect of GABA against ischemia/reperfusion-induced renal injury in rats. J Pharmacol Exp Ther 2011;338:767–774.
39 Li C, Zhang D, Han S, et al: Synthesis, electronic properties, and applications of indium oxide nanowires. Ann NY Acad Sci 2003;1006:104–121.
40 Singh K, Xiao L, Remondino A, Sawyer DB, Colucci WS: Adrenergic regulation of cardiac myocyte apoptosis. J Cell Physiol 2001;189: 257–265.
41 Kim S, Iwao H: Molecular and cellular mechanisms of angiotensin II-mediated cardiovascular and renal diseases. Pharmacol Rev 2000; 52:11–34.
42 Kajstura J, Cigola E, Malhotra A, et al: Angiotensin II induces apoptosis of adult ventricular myocytes in vitro. J Mol Cell Cardiol 1997;29:859–870.
43 Kawano H, Do YS, Kawano Y, et al: Angiotensin II has multiple profibrotic effects in human cardiac fibroblasts. Circulation 2000; 101:1130–1137.
44 Desai KV, Laine GA, Stewart RH, et al: Mechanics of the left ventricular myocardial interstitium: effects of acute and chronic myocardial edema. Am J Physiol Heart Circ Physiol 2008;294:H2428–H2434.
45 Nath KA, Grande JP, Croatt AJ, et al: Transgenic sickle mice are markedly sensitive to renal ischemia-reperfusion injury. Am J Pathol 2005;166:963–972.
46 Robinson SC, Bowmer CJ, Yates MS: Cardiac function in rats with acute renal failure. J Pharm Pharmacol 1992;44:1007–1014.

47 Kelly KJ, Meehan SM, Colvin RB, Williams WW, Bonventre JV: Protection from toxicant-mediated renal injury in the rat with anti-CD54 antibody. Kidney International 1999;56:922–931.

48 Wencker D, Chandra M, Nguyen K, et al: A mechanistic role for cardiac myocyte apoptosis in heart failure. J Clin Invest 2003;111: 1497–1504.

49 Le NT, Richardson DR: The role of iron in cell cycle progression and the proliferation of neoplastic cells. Biochim Biophys Acta 2002; 1603:31–46.

50 Liu YH, D'Ambrosio M, Liao TD, et al: N-acetyl-seryl-aspartyl-lysyl-proline prevents cardiac remodeling and dysfunction induced by galectin-3, a mammalian adhesion/growth-regulatory lectin. Am J Physiol Heart Circ Physiol 2009;296:H404–H412.

51 Wiedemann HP, Wheeler AP, Bernard GR, et al: Comparison of two fluid-management strategies in acute lung injury. N Engl J Med 2006;354:2564–2575.

52 Arikan AA, Zappitelli M, Goldstein SL, Naipaul A, Jefferson LS, Loftis LL: Fluid overload is associated with impaired oxygenation and morbidity in critically ill children. Pediatr Crit Care Med 2012;13:253–258.

53 Bouchard J, Soroko SB, Chertow GM, et al: Fluid accumulation, survival and recovery of kidney function in critically ill patients with acute kidney injury. Kidney Int 2009;76:422–427.

54 Payen D, de Pont AC, Sakr Y, Spies C, Reinhart K, Vincent JL: A positive fluid balance is associated with a worse outcome in patients with acute renal failure. Crit Care 2008; 12:R74.

55 Grams ME, Estrella MM, Coresh J, Brower RG, Liu KD: Fluid balance, diuretic use, and mortality in acute kidney injury. Clin J Am Soc Nephrol 2011;6:966–973.

56 Bellomo R, Cass A, Cole L, et al, RENAL Replacement Therapy Study Investigators: An observational study fluid balance and patient outcomes in the Randomized Evaluation of Normal vs. Augmented Level of Replacement Therapy trial. Crit Care Med 2012;40:1753–1760.

57 Ip JE, Cheung JW, Park W: Temporal associations between thoracic volume overload and malignant ventricular arrhythmias: a study of intrathoracic impedance. J Cardiovasc Electrophysiol 2011;22:293–299.

58 Davis KL, Laine GA, Geissler HJ: Effects of myocardial edema on the development of myocardial interstitial fibrosis. Microcirculation 2000;7:269–280.

59 Nazar A, Guevara M, Sitges M, et al: LEFT ventricular function assessed by echocardiography in cirrhosis: relationship to systemic hemodynamics and renal dysfunction. J Hepatol 2013;58:51–57.

60 Legrand M, Darmon M, Joannidis M, Payen D: Management of renal replacement therapy in ICU patients: an international survey. Intensive Care Med 2013;39:101–108.

61 Bellomo R, Cass A, Cole L, et al: Intensity of continuous renal-replacement therapy in critically ill patients. N Engl J Med 2009;361: 1627–1638.

62 Nimmo AJ, Than N, Orchard CH, Whitaker EM: The effect of acidosis on β-adrenergic receptors in ferret cardiac muscle. Exp Physiol 1993;78:95–103.

63 Bagshaw SM, Wald R, Barton J, et al: Clinical factors associated with initiation of renal replacement therapy in critically ill patients with acute kidney injury – a prospective multicenter observational study. J Crit Care 2012;27:268–275.

64 Palevsky PM, Zhang JH, O'Connor TZ, et al: Intensity of renal support in critically ill patients with acute kidney injury. N Engl J Med 2008;359:7–20.

65 Schwenger V, Weigand MA, Hoffmann O, et al: Sustained low efficiency dialysis using a single-pass batch system in acute kidney injury – a randomized interventional trial: the REnal Replacement Therapy Study in Intensive Care Unit PatiEnts. Crit Care 2012; 16:R140.

66 Zazzo JF, Troche G, Ruel P, Maintenant J: High incidence of hypophosphatemia in surgical intensive care patients: efficacy of phosphorus therapy on myocardial function. Intensive Care Med 1995;21:826–831.

67 Demirjian S, Teo BW, Guzman JA, et al: Hypophosphatemia during continuous hemodialysis is associated with prolonged respiratory failure in patients with acute kidney injury. Nephrol Dial Transplant 2011; 26:3508–3514.

68 Duranton F, Cohen G, De Smet R, et al: Normal and pathologic concentrations of uremic toxins. J Am Soc Nephrol 2012;23: 1258–1270.

69 Neirynck N, Vanholder R, Schepers E, Eloot S, Pletinck A, Glorieux G: An update on uremic toxins. Int Urol Nephrol 2013;45: 139–150.
70 Littmann L, Monroe MH, Taylor L 3rd, Brearley WD Jr: The hyperkalemic Brugada sign. J Electrocardiol 2007;40:53–59.
71 Marques JS, Diogo AN: Dead man walking: an extreme case of sinusoidal wave pattern in severe hyperkalemia. J Am Coll Cardiol 2012; 59:2118.
72 Shannon TR, Pogwizd SM, Bers DM: Elevated sarcoplasmic reticulum Ca^{2+} leak in intact ventricular myocytes from rabbits in heart failure. Circ Res 2003;93:592–594.
73 Schulz R, Panas DL, Catena R, Moncada S, Olley PM, Lopaschuk GD: The role of nitric oxide in cardiac depression induced by interleukin-1β and tumour necrosis factor-α. Br J Pharmacol 1995;114:27–34.
74 Tune JD, Gorman MW, Feigl EO: Matching coronary blood flow to myocardial oxygen consumption. J Appl Physiol 2004;97:404–415.
75 Ronco C, Cruz D, Noland BW: Neutrophil gelatinase-associated lipocalin curve and neutrophil gelatinase-associated lipocalin extended-range assay: a new biomarker approach in the early diagnosis of acute kidney injury and cardio-renal syndrome. Seminars in nephrology 2012;32:121–128.
76 Mishra J, Ma Q, Prada A, et al: Identification of neutrophil gelatinase-associated lipocalin as a novel early urinary biomarker for ischemic renal injury. J Am Soc Nephrol 2003;14:2534–2543.
77 Haase M, Haase-Fielitz A, Bellomo R, Mertens PR: Neutrophil gelatinase-associated lipocalin as a marker of acute renal disease. Curr Opin Hematol 2010.
78 Schmidt-Ott KM, Mori K, Li JY, et al: Dual action of neutrophil gelatinase-associated lipocalin. J Am Soc Nephrol 2007;18:407–413.
79 Cruz DN, Gaiao S, Maisel A, Ronco C, Devarajan P: Neutrophil gelatinase-associated lipocalin as a biomarker of cardiovascular disease: a systematic review. Clin Chem Lab Medicine 2012;50:1533–1545.
80 Yndestad A, Landro L, Ueland T, et al: Increased systemic and myocardial expression of neutrophil gelatinase-associated lipocalin in clinical and experimental heart failure. Eur Heart J 2009; 30:1229–1236.
81 Ding L, Hanawa H, Ota Y, et al: Lipocalin-2/neutrophil gelatinase-B associated lipocalin is strongly induced in hearts of rats with autoimmune myocarditis and in human myocarditis. Circ J 2010;74:523–530.
82 Carlsson AC, Larsson A, Helmersson-Karlqvist J, et al: Urinary kidney injury molecule 1 and incidence of heart failure in elderly men. Eur J Heart Fail 2012 (E-pub ahead of print).
83 Scheuer J, Stezoski W: The effects of uremic compounds on cardiac function and metabolism. J Mol Cell Cardiol 1973;5:287–300.
84 James MT, Ghali WA, Knudtson ML, et al: Associations between acute kidney injury and cardiovascular and renal outcomes after coronary angiography. Circulation 2011;123: 409–416.
85 Selby NM, Kolhe NV, McIntyre CW, et al: Defining the cause of death in hospitalised patients with acute kidney injury. PLoS One 2012;7:e48580.
86 Thakar CV, Christianson A, Himmelfarb J, Leonard AC: Acute kidney injury episodes and chronic kidney disease risk in diabetes mellitus. Clin J Am Soc Nephrol 2011;6: 2567–2572.
87 James MT, Hemmelgarn BR, Wiebe N, et al: Glomerular filtration rate, proteinuria, and the incidence and consequences of acute kidney injury: a cohort study. Lancet 2010; 376:2096–2103.
88 Acute Dialysis Quality Initiative (ADQI). www.ADQI.org (accessed January 10, 2013).

Dr. Sean M. Bagshaw
Division of Critical Care Medicine, University of Alberta Hospital
3C1.12 Walter C. Mackenzie Centre
8440-122 Street, Edmonton, AB T6G2B7 (Canada)
E-Mail bagshaw@ualberta.ca

McCullough PA, Kellum JA, Mehta RL, Murray PT, Ronco C (eds): ADQI Consensus on AKI Biomarkers and Cardiorenal Syndromes. Contrib Nephrol. Basel, Karger, 2013, vol 182, pp 158–173
DOI: 10.1159/000349972

Cardiorenal Syndrome Type 4: Insights on Clinical Presentation and Pathophysiology from the Eleventh Consensus Conference of the Acute Dialysis Quality Initiative (ADQI)

James A. Tumlin[a] · Maria R. Costanzo[b] · Lakhmir S. Chawla[c] · Charles A. Herzog[d] · John A. Kellum[e] · Peter A. McCullough[f] · Claudio Ronco[g] · for the Acute Dialysis Quality Initiative (ADQI) 11 Consensus Group

[a]Department of Medicine, College of Medicine, University of Tennessee, Chattanooga, Tenn., USA; [b]Midwest Heart Foundation, Edward Heart Hospital, Naperville, Ill., USA; [c]Department of Anesthesiology and Critical Care Medicine and Division of Renal Diseases and Hypertension, Department of Medicine, George Washington University Medical Center, Washington, D.C., USA; [d]Department of Internal Medicine, Chronic Disease Research Group, Hennepin County Medical Center, University of Minnesota, Minneapolis, Minn., USA; [e]Clinical Research, Investigation, and Systems Modeling of Acute Illness Center, Department of Critical Care Medicine, University of Pittsburgh, Pittsburgh, Pa., USA; [f]St. John Providence Health System, Warren Mich., Providence Hospitals and Medical Centers, Southfield and Novi, Mich., St. John Macomb Oakland Center, Madison Heights, Mich., St. John Hospital and Medical Center, Detroit, Mich., USA; [g]Department of Nephrology, Dialysis and Transplantation and International Renal Research Institute (IRRIV), San Bortolo Hospital, Vicenza, Italy

Abstract

In developed countries, the continuing rise in the prevalence of hypertension, hyperlipidemia and diabetes has contributed to an overall increase in the incidence of both cardiovascular disease (CVD) and chronic kidney disease (CKD). The observation that even modest reductions in renal function correlate with increased CVD morbidity and mortality has led to the recognition that CKD is an independent risk factor for CVD. Conversely, there is a growing recognition that many pathologic conditions that contribute to CVD, including coronary artery disease, left ventricular hypertrophy and diastolic dysfunction, can accelerate the decline in renal function. In addition, physiologic mechanisms designed to compensate for reduced glomerular filtration rate including activation of the

ADQI 11 Workgroup members are listed in Appendix 2.

renin-angiotensin-aldosterone axis, the release of fibroblastic growth factor 23 and other mechanisms for calcium-phosphate homeostasis as well as and the pathophysiologic effects of uremic toxins can also directly contribute to CVD. The end result of the interaction between changes in pressure and volume overload and the physiologic compensation for the loss of function in both the heart and the kidney leads to accelerated decline in both organ systems. This complex physiologic and pathophysiologic interplay between the cardiovascular and renal systems is collectively referred to as the cardiorenal syndrome. The discussion which follows is aimed at outlining the pathophysiologic mechanisms linking advanced CKD (4 and 5) to the development of cardiac abnormalities which occur with unique frequency and severity in patients with severe impairment of renal function.

Epidemiology of Cardiovascular Disease in Chronic Kidney Disease

The rising prevalence of chronic kidney disease (CKD) and end-stage renal disease (ESRD) is a global medical and epidemiological problem [1]. In the United States it is estimated that up to 13% of the population (30 million people) have CKD. A growing body of evidence shows that declining renal function is an independent risk for cardiovascular disease (CVD). Patients with CKD are at higher risk for death following myocardial infarction and patients experiencing even transient renal dysfunction have increased long-term risk for CVD [1, 2].

While the mechanisms leading to increased risks for cardiovascular complications among CKD patients are not fully known, there is a growing awareness of the interdependence between these two organ systems. The renal response to impaired glomerular filtration rate (GFR) can lead to activation of multiple compensatory pathways including upregulation of the renin-angiotensin-aldosterone axis (RAAS) and sympathetic nervous system as well as activation of the calcium-parathyroid axis. These physiologic responses can be due to underlying diseases such as hypertension or diabetes or can be a response to the functional decline of either the heart or the kidney. The loss of renal mass leads to the accumulation of total body sodium and water with the subsequent stimulation of angiotensin and aldosterone production. The resulting hypertension coupled with direct effects of angiotensin and aldosterone on cardiac myocytes accelerates left ventricular hypertrophy (LVH) and cardiac fibrosis. The cardiorenal syndrome collectively describes the complex interactions between the physiologic and pathophysiologic consequences of declining function in both the heart and the kidney. The ADQI group has created separate classifications for the cardiorenal syndrome in an attempt to isolate some of the unique pathophysiologic mechanisms of this disorder with the intent to devise new strategies of treatment [3].

Coronary Atherosclerotic Heart Disease and Chronic Kidney Disease

Numerous epidemiologic and clinical studies have documented the association between even modest reductions in renal function and the increased risk for cardiovascular death. While there is a clear inverse association between reduced GFR and cardiovascular morbidity, the mechanisms that lead to an increased incidence of CVD following the onset of CKD are poorly understood. Patients with CKD have increased rates of atherosclerotic coronary artery disease (CAD), myocardial infarctions, LVH, and have an increased risk for sudden death. Longitudinal studies demonstrate that the risk for cardiovascular complications increases with progressive loss of renal function. The risks for cardiovascular complications for patients with eGFRs <30 ml/min/1.73 m^2 are up to 10-fold higher than those with eGFRs >60 ml/min/1.73 m^2 [4]. This startling prevalence of CVD complications exceeds the risk expected from typical cardiovascular risk factors (hypertension, hyperlipidemia, etc.) and suggests that the loss of renal function may directly contribute to the development of cardiovascular complications [5–7]. Small limited studies suggest that patients with early as well as late stage CKD have a higher prevalence of angiographically significant CAD. For example, Joki et al. [8] performed coronary angiography in 24 pre-dialysis ESRD patients in order to determine the prevalence and severity of coronary disease. When clinically significant coronary disease was defined as greater than 75% of one or more coronary vessels, Joki et al. found that 73% of symptomatic and 54% of asymptomatic CKD stage V patients had significant CAD. Moreover, these patients were more likely to have multivessel disease and ECG evidence of prior of ischemia. The high prevalence of CAD among asymptomatic CKD patients has been noted by other investigators. Ohtake et al. [9] performed coronary angiograms in 30 asymptomatic and 39 symptomatic CKD patients and found that 53% of asymptomatic patients had significant CAD. Of these patients, 10 (62.5%) had single-vessel disease, 4 (25%) had two-vessel disease, and 2 (12.5%) had triple-vessel disease. Moreover, 31% of these patients had greater than 90% luminal narrowing. It is also interesting to note that when stress nuclear cardiac scintigraphy was performed in these patients only 40% were found to have significant perfusion defects. This finding raised the question of whether stress nuclear cardiac scintigraphy is the appropriate method for CAD screening within the CKD population [9, 10]. A recent analysis from the Cochrane investigators demonstrated that the best accuracy for non-invasive screening for CAD in renal transplant candidates is achieved with dobutamine stress echocardiography [11]. To determine whether increased rates of CAD are also present in patients with less advanced CKD, Chonchol et al. [12] reviewed the cardiac catheterization studies in 261 patients with eGFR between 30 and 90 ml/min. Despite having relative preserved renal function, 51% of patients with eGFR >90 ml/min had a 70% stenosis in at least one coronary artery. In contrast, more than 84% patients with advanced renal disease (eGFR <30) were found to have significant CAD. Moreover, patients with advanced CKD were more likely to have significant left main and multivessel disease. These observations were supported by Charytan et al. [13] who also noted that compared to normal controls, patients with CKD and acute coronary syndrome were more likely to have atherosclerotic obstructions in the proximal portion of the coronary arteries [14].

Accelerated Coronary Calcification, Myocardial Calcification, and Reduced Aortic Compliance in CKD

The prevalence of atherosclerotic heart disease among CKD/ESRD patients is not sufficient to explain the observed rates of sudden cardiac death and cardiovascular complications [15]. While the mechanism(s) linking CKD to the occurrence of CVD events are not fully understood, osteoblastic transformation of vascular smooth muscle cells appears to be a critical component of the pathogenesis of vascular calcification in patients with CKD. There is a growing recognition that altered calcium-phosphate metabolism, impaired vitamin D production and secondary hyperparathyroidism can directly contribute to vascular calcification as a result of their effects on these osteoblastic-like cells. Hydroxyapatite is the principal structural component of bone and is produced by osteoblastic cells in both endothelial atheromata and the vascular media. These pathophysiologic mechanisms are hallmarks of the uremic condition and help to explain the prevalence of vascular calcification among CKD/ESRD patients [16]. The presence of coronary artery calcification is a potent predictor of cardiac events and may contribute to reduced coronary flow reserve observed in CKD/ESRD patients [17, 18]. The risk for developing coronary calcification increases with advancing renal disease. In a study of 119 pre-dialysis CKD patients, Garland et al. [19] quantified coronary calcification using high-resolution, multislice computed tomography and demonstrated that calcification of coronary vessels was detectable as early as CKD stage III. Of the patients studied, 83% had some degree of coronary calcification with a full 32% exhibiting severe disease. Interestingly, the severity and prevalence of calcification did not correlate with the stage of CKD. The development of vascular calcifications is not limited to coronary vessels but has also been detected in peripheral arteries. This observation may explain the higher rates of lower extremity amputations among CKD/ESRD patients [20]. In contrast to non-CKD patients, the quantity and density of calcium deposits is greater among CKD patients. Recent studies using electron beam scanning and high-resolution computed tomography scans consistently found calcium deposits of several millimeters in depth raising the question of whether calcium deposits are not limited to the intima, but also extend into the vessel media. To investigate this question, Moe et al. [21] examined the inferior epigastric arteries in 41 ESRD patients undergoing renal transplantation and confirmed that arterial calcification was common in the medial layer of the vessel wall. In addition, immunostaining of calcified areas demonstrated the presence of bone matrix proteins including osteopontin, type I collagen and bone sialoprotein. In an autopsy study, Gross et al. [17] compared the distribution of coronary calcification in uremic and non-uremic patients and demonstrated that medial calcification was present in 16% of uremic patients but only 3% of those without renal dysfunction. Interestingly, medial calcification was associated with the presence of bone matrix proteins including osteocalcin as well as markers of chronic inflammation including transforming growth factor-β (TGF-β) and components of activated complement [22].

Increased calcium content of the myocardium has been demonstrated in ESRD patients and may contribute to reduced left ventricular (LV) compliance. It is also possible that disordered movement of calcium and potentially other electrolytes managed by myocytes and conducting tissue is responsible for the variety and prevalence of arrhythmias in CKD and ESRD. Aortic calcification is associated with reduced compliance as demonstrated on multiple studies [23, 24] Thus, the shear stresses with each systole are greater in patients with CKD and ESRD, and in theory, have greater degrees of end-organ injury due to hypertension compared to those with more compliant vessels [24]. Moreover, even in the absence of epicardial coronary disease, reduced aortic compliance reduces coronary artery perfusion. While the mechanism is not fully understood, a generally accepted mechanistic view is that with falling aortic compliance reflected pulse waves in late systole raise central pressure which induces subendocardial ischemia through reduced diastolic filling [25].

Left Ventricular Hypertrophy and Chronic Kidney Disease

Both LVH and reduced LV function have long been recognized as complications of advanced renal disease. Moreover, the prevalence of LVH increases with declining renal function. Traditional risk factors including hypertension, diabetes mellitus, and volume overload contribute to the development of LVH in CKD populations. The observation that hyperparathyroidism, malnutrition, retained products of uremia and even the delivery of dialysis can contribute to

LVH has raised the question of whether CKD/ESRD populations are uniquely predisposed to the development of CVD. For example, the prevalence of LVH varies from 16 to 31% in individuals with GFR >30 ml/min, increasing to 60–75% prior to starting renal replacement therapy, and up to 90% of patients after the initiation of dialysis [23]. In a prospective study of 433 ESRD patients, Foley et al. [26] noted that 74% of patients had echocardiographic evidence of LVH with over 30% having concurrent LV failure. In a subsequent study, Foley et al. [27] prospectively followed 596 incident hemodialysis patients with no prior history of cardiac disease and determined whether the incidence of LVH correlates with duration of dialysis. Using serial echocardiograms to measure left ventricular mass index (LVMI), Foley et al. demonstrated that after 18 months of dialysis, LVMI increased in 62% patients with 49% developing overt LV failure. These observations raise the question of whether the very process of dialysis facilitates the development of LVH in ESRD patients. However, the more comprehensive and detailed IDEAL study failed to demonstrate that the initiation of dialysis in stage V CKD patients worsened existing LVH, suggesting that dialysis alone does not contribute to LVH nor is it able to reverse long-standing hypertrophic changes [13, 28]. The mechanisms which contribute to the 'cardiomyopathy of advanced CKD' are unknown, but a growing body of evidence indicates that retained products of uremia can directly alter cardiac structure. Because many uremic toxins are non-polar and highly protein-bound, their clearance by conventional dialyzers is very limited. These observations suggest that while dialysis itself may not directly contribute to the LVH associated with CKD, its inability to clear specific uremic toxins could indirectly contribute to cardiac remodeling. This is not to say that dialysis cannot alter cardiac function. For example, the elegant work of McIntyre et al. [29] has shown that conventional hemodialysis in children and adults may cause regional myocardial stunning through a mechanism that does not correspond to coronary artery vascular territories. This observation suggests that the wall motion abnormalities observed during dialysis are a form of myocardial stunning and should be considered as a form of non-Takosubo-type stress cardiomyopathy.

The clinical conditions that contribute to the development of LVH in CKD are similar to those occurring in other patient cohorts and include hypertension, pressure overload, and activation of the RAAS. Arteriosclerosis and hypertension promote myocyte hypertrophy which results in an increased LV mass through myocyte hypertrophy, an increased wall thickness and ultimately reduced ventricular compliance. Pathologic adaptations to increasing wall thickness contribute to a loss of capillary density, secondary myocardial fibrosis and further compensatory hypertrophy [30]. Clinical conditions that contribute to

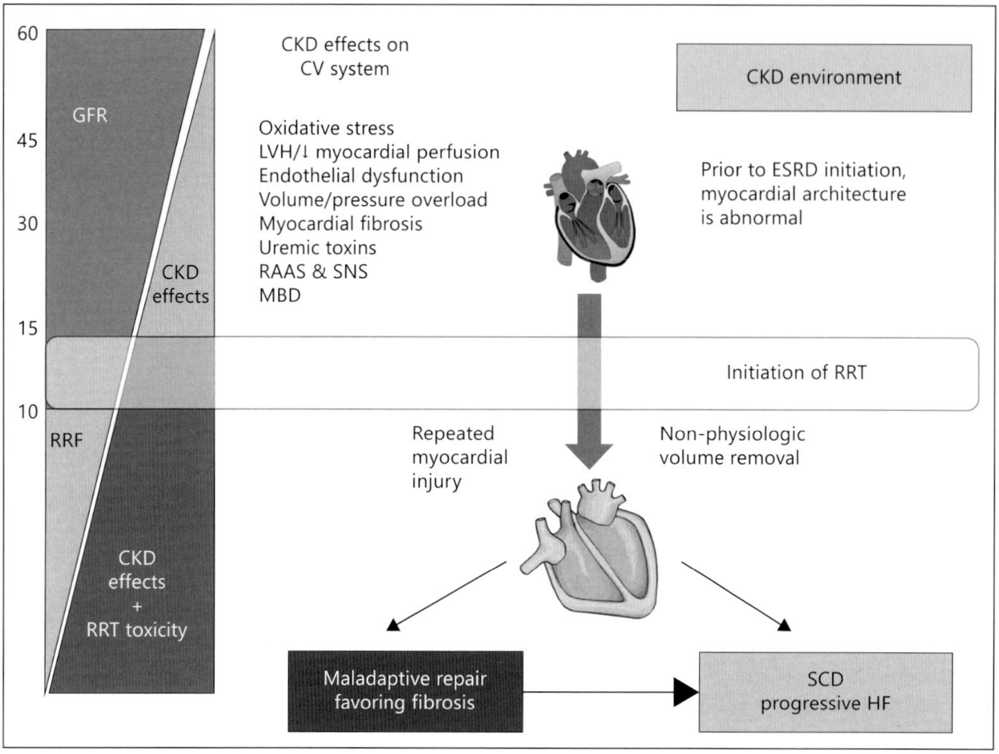

Fig. 1. Declining GFR is associated with multiple stressors upon the CV system including volume/pressure overload, oxidative stress, and activation of the RAAS and neurohumoral pathways. Prior to the initiation of renal replacement therapy, the heart undergoes maladaptive responses including reduced diastolic compliance, LVH, reduction in myocardial capillary density and uremia-induced myocardial fibrosis. Following the initiation of renal replacement therapy, 'non-physiologic' fluid removal worsens myocardial ischemia leading to progressive heart failure and sudden cardiac death.

LVH in the general population such a reduced aortic compliance are often more common and severe in CKD patients. For example, arteriosclerotic changes that reduce aortic compliance in the general population are also present in CKD patients, but other variables including hyperphosphatemia and secondary vascular calcifications can further affect aortic compliance [31]. Moreover, the progressive loss of nephron mass inherent to CKD leads to accumulation of salt and water which secondarily contributes to hypertension and pressure and volume overload. These same conditions contribute to LVH through upregulation of RAAS subsequent release of pro-fibrotic factors such as galectin-3, TGF-β and endogenous cardiac steroids [32]. The unique relationship between CKD and the development of cardiovascular complications is supported by the beneficial effects of renal transplantation on cardiac structure and remodeling (fig. 1, 2).

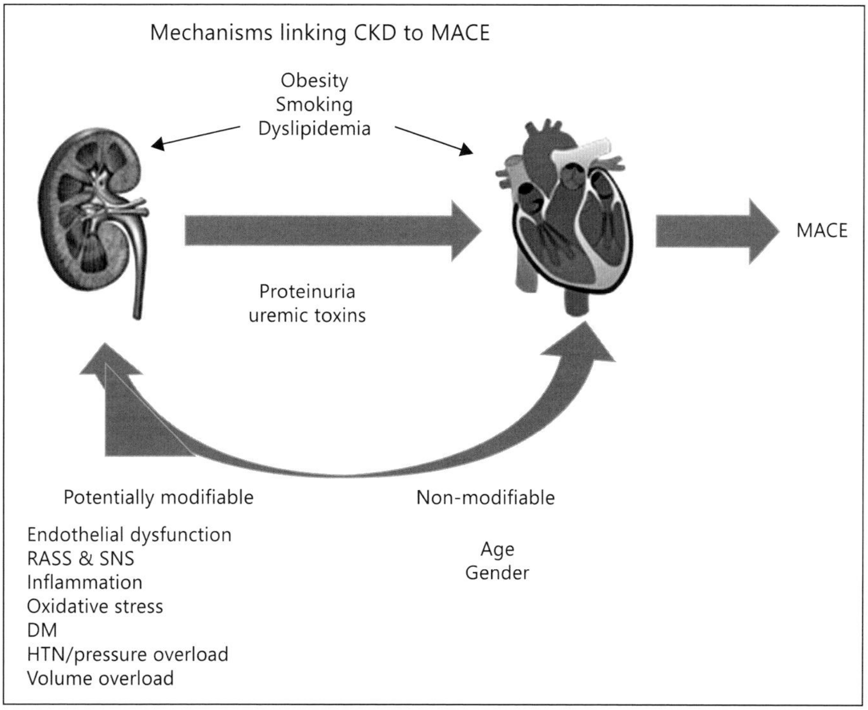

Fig. 2. Modifiable and non-modifiable variables contribute to the link between CKD and major adverse cardiovascular events. Age and gender are non-modifiable variables, hypertension, correction of dyslipidemia, activation of RAAS and improved clearance of uremic toxins could be modified over time and reduce the prevalence of CRS-4.

For example, Wali et al. [33] studied 138 patients being evaluated for renal transplantation with established LV dysfunction defined as LVEF ≤40% that was not correctable by revascularization. Interestingly, 69% of patients demonstrated a post-transplant increase in LVEF from a mean of 32 to 47% at 6 months and 57% 1 year after transplantation. These cardiovascular improvements were also shown to translate into improved patient survival. Patients whose post-transplant LVEF improved to greater than 50% were found to have an all-cause mortality that was 8-fold lower than patients whose LVEF failed to normalize.

Cardiac Myocyte-Capillary Mismatch

A complication of hypertrophic cardiomyopathies is that as cardiac myocytes enlarge, the density of capillaries needed to accommodate the increased oxygen demand is progressively reduced over time. To determine whether this effect is exacerbated by uremic conditions, Amann et al. [34] compared myocardial capillary density in normal controls with dialysis patients and those with hypertensive cardiomyopathies. While hypertensive patients demonstrated a reduction capillary density, this decrease was even more pronounced in ESRD patients and nearly 2-fold lower than normal controls. Moreover, myocyte diameter and volume of the interstitial space were significantly increased in uremic patients compared to the other groups. Amann et al. [34] observed that the collagen content of the myocardial interstitium rose to 25% – a value that is observed in patients with severe aortic stenosis and other conditions of pressure overload. Prolonged periods of hemodynamic stress can induce cardiac remodeling which includes the increased expression of interstitial myofibroblasts – a cell type that is not present in normal myocardium and is characterized by its fibrogenic potential. The observation that ESRD patients develop clinically significant reductions in myocardial capillary density raises the question of whether this observed histopathology is unique to the uremic condition [34]. On a clinical level, a reduction in the capillary density of the uremic myocardium may explain in part the exaggerated intolerance of ESRD patients to ischemia as well as the higher prevalence of diastolic dysfunction [35]. The combination of intolerance to myocardial ischemia, LVH, and myocardial fibrosis, coupled with non-physiologic alterations in volume and electrolytes (particularly in CKD-5 hemodialysis patients), renders ESRD patients particularly vulnerable to sudden cardiac death.

Uremia and Cardiac Fibrosis

In addition to hypertrophic changes in the heart, a growing body of evidence suggests that patients with advanced CKD develop a unique form of cardiac and fibrosis [36, 37]. For example, early post-mortem studies by Mall et al. [37] describe a unique form of cardiac and renal fibrosis that was observed in over 90% of CKD patients, but absence in non-uremic controls. Both ESRD patients and those with pre-dialysis CKD demonstrated a uniform pattern of intermyocardial fibrosis which differs from the endocardial to epicardial perivascular fibrosis pattern that is typical of hypertensive and ischemic cardiomyopathies. By light microscopy, trichrome staining and light microscopy consistently demonstrated diffuse interstitial fibrosis, while electron microscopy showed the presence of collagen but an absence of P2-M amyloid protein [37]. Figure 3a is representative of normal myocardium without interstitial fibrosis and a typical myocyte to capillary ratio. Figure 3b is representative of a patient with hypertensive cardiomyopathy. While individual myocytes are hypertrophied, myocar-

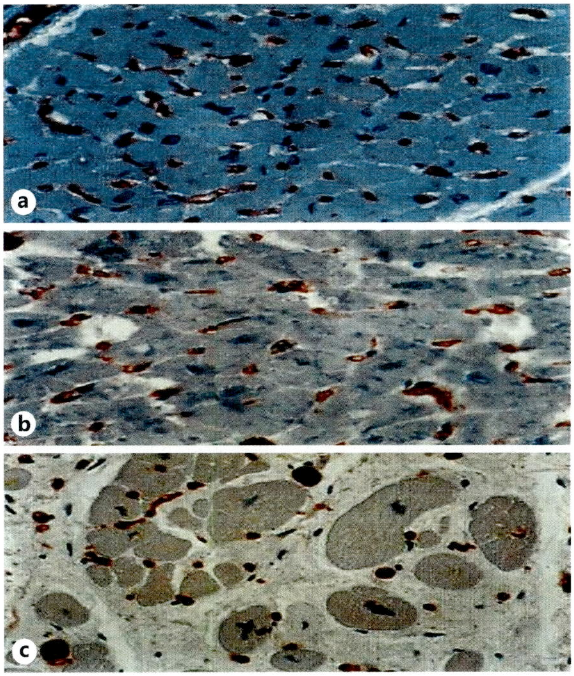

Fig. 3. a Myocardial biopsy of control patient. Size of myocytes estimated between vessels (red). ×200. **b** Myocardium of hypertensive patient: increased myocyte cross-sectional area; reduced number capillaries per unit area; increased myocardial interstitial tissue. ×200. **c** Myocardium of CKD patient: myocardial cross section markedly increased. Further reduction in myocardial capillary density. Reproduced with permission from [37].

dial fibrosis and capillary density are normal. This is contrasted by cardiac tissue obtained from a patient with CKD (fig. 3c). Of note, there is increased deposition of collagen and other extracellular matrix proteins with an associated reduction in the capillary/myocyte ratio [34]. The mechanisms leading to CKD/ESRD cardiac fibrosis are poorly understood, but recent studies suggest that substances in the plasma of uremic patients' components of uremia plasma including indoxyl sulfate and *p*-cresol can directly contribute to cardiac fibrosis. Indoxyl sulfate and other non-polar uremic toxins are highly protein-bound and thus exhibit very poor clearance by dialysis. While indoxyl sulfate concentrations in healthy subjects range between 1 and 2.9 mM, they can exceed 500 mM in patients with advanced CKD [28]. At these concentrations, indoxyl sulfate directly contributes to the generation of cardiac fibrosis through the synthesis of TGF-β, tissue inhibitor of metalloproteinase-1 (TIMP-1) and α1 collagen. Moreover, protein-bound uremic toxins have been shown to stimulate cardiac remodeling through activation of the p38 MAPK, p42/44 MAPK, and NF-κB pathways. These observations underscore the need to identify the clinical consequences of the cardiorenal syndrome and to define the unique physiologic and pathophysiologic interdependence between the two organs [38–40].

Galectin-3 and Cardiac Remodeling

There is an increasing recognition that diastolic dysfunction and other manifestations of decompensated heart failure are associated with myocardial infiltration by macrophages and monocytes. While a number of candidate mediators for these changes have been investigated, recent studies have consistently demonstrated a marked upregulation of galectin-3 [41]. Galectin-3 is a member of the β-galactoside-binding lectin family that is synthesized by macrophages and known to interact with specific extracellular matrix proteins including laminin, synexin and integrins. Pre-clinical studies find that galectin-3 can also bind directly to cardiac fibroblasts and reduced LV function through an increase in collagen production [42]. In a recent study of 232 patients with NYHA class III or IV heart failure, Lok et al. [43] used NT-proBNP and eGFR to adjust for severity of heart disease and degree of renal dysfunction and demonstrated that serum galectin-3 levels were independent predictors of cardiovascular mortality. Moreover, Lin et al. [44] followed 106 patients with NYHA class II heart failure and found a direct correlation between serum galectin-3 levels and markers of chronic inflammation. The observation that galectin-3 levels correlate with type III amino-terminal propeptide of procollagen (PIIINP), MMP-2, and TIMP-1 suggests that myocardial macrophage infiltration enhances turnover of extracellular matrix proteins in patients with established heart failure. Despite the role of infiltrating macrophages in the synthesis of galectin-3, studies have failed to show a correlation between the severity of heart failure and plasma galectin-3 levels. Gopal et al. [45] studied 119 patients with varying degrees of heart failure and failed to correlate galectin-3 levels with LV ejection fraction or level of cardiac function. In contrast, plasma galectin-3 levels are significantly affected by changes in renal clearance. Gopal et al. noted that for patients with an eGFR <30 ml/min, plasma galectin-3 levels can be 2-fold higher than normal controls.

Fibroblast Growth Factor 23

The link between compensation for falling GFR and the subsequent downstream effects of cardiovascular structure and function is a central component of the cardiorenal syndrome (fig. 4). Recent pre-clinical and clinical studies

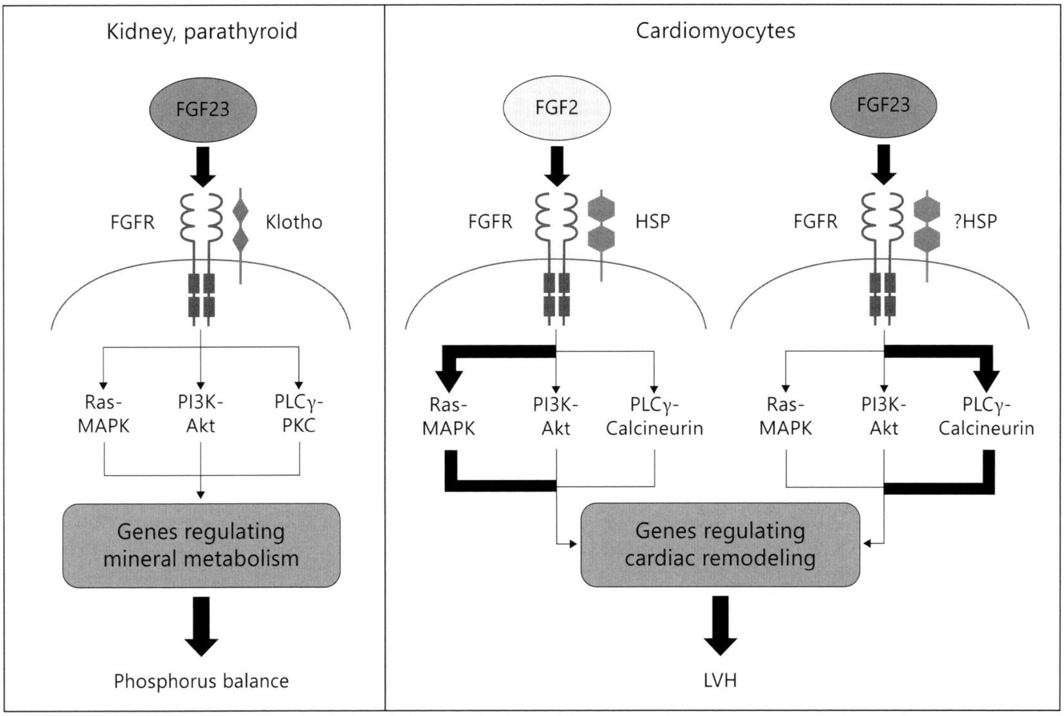

Fig. 4. In the kidney and parathyroid glands, FGF23 signaling requires FGFR and the co-receptor klotho. FGF23-klotho binding to FGFR stimulates autophosphorylation of the receptor tyrosine kinase and induces signaling through three major pathways: Ras-MAPK, PI3K-Akt, and PLC-PKC. FGF23 regulates phosphorus balance by altering expression of genes involved in parathyroid, vitamin D, and phosphorus metabolism. In cardiomyocytes, FGF2 signaling requires FGFR and heparan sulfate proteoglycans (HSP) as co-receptor and signals primarily through the Ras-MAPK pathway. Binding of FGF23 to FGFR on cardiomyocytes stimulates autophosphorylation of the receptor tyrosine kinase independent of klotho, which is not expressed in cardiomyocytes, and signals primarily through the PLC-calcineurin pathway. Whether HSP acts as co-receptor remains to be determined. Reproduced with permission from [46].

have shown an association between the family of fibroblast growth factors (FGF) and the development of a pathologic and load-independent form of LVH. The FGFs are a family of peptides with broad biologic functions that include the regulation of growth and differentiation of cardiac myocytes. In contrast to other members of the FGF family, FGF23 has evolved unique paracrine functions in the kidney which facilitate phosphate excretion by blocking the synthesis of vitamin D_3 and inhibition of phosphate resorption in the proximal nephron [46]. The loss of nephron mass and subsequent accumulation of phosphate leads to a compensatory rise in FGF23 levels. While this physiologic compensation serves to maintain calcium-phosphate metabolism, prolonged elevations of FGF23 can contribute to LVH and cardiac remodeling. Recent studies have shown that even modest reductions in GFR can stimulate FGF23 production. Gutiérrez et al. [47] noted that for patients with CKD stage II, FGF23 levels can be 2–5 times above normal values, while levels in ESRD patient are often a 1,000-fold higher than normal controls [48]. In an early attempt to link FGF23 levels development of LVH, Gutiérrez et al. [47] used echocardiography to determine LVMI and coronary artery calcification in 162 patients with CKD. Using multivariable-adjusted regression analysis, the authors noted a 5% rise in LVMI for every log increase in FGF23 levels. Moreover, patients in the highest tertile of FGF23 were found to have a 2.4-fold higher risk for coronary artery calcifications. While the data from this study and others like it were suggestive of a causal role for FGF23 in CVD of CKD patients, the studies were limited by small sample size and cross-sectional design [49]. More recently, Faul et al. [46] utilized a large multicenter prospective study cohort to examine the potential causal relationship between FGF23 and the development of LVH in CKD patients with GFR between 20 and 70 ml/min. In the highest quartile of FGF23, the prevalence of concentric LVH was 49% while evidence of cardiac remodeling was seen in 20% of patients.

Conclusions

The complexity of cardiorenal syndrome (CRS) along with its associated comorbidities poses a significant challenge for the cardiology and nephrology communities. The ADQI group believes that prior to the creation of meaningful options for the prevention and treatment of this disorder, it will be necessary to identify subgroups of CRS patients that are defined by their underlying pathophysiology. We have defined the CRS-4 patients by focusing upon both the primary and compensatory physiologic responses to declining organ function that contributes to clinical phenotype of CRS-4. We highlighted specific clinical characteristics of CRS-4 that are critical to patient diagnosis and clinical management including accelerated coronary disease, cardiac hypertrophy/fibrosis and the direct role of dialysis in cardiovascular outcomes. We propose that CRS-4 is a unique subgroup of the CRS and that by defining this specific group of patients, it will enable the creation of disease-specific therapies.

References

1 Redón J, Cea-Calvo L, Lozano JV, Fernández-Pérez C, Navarro J, Bonet A, González-Esteban J, ERIC-HTA 2003 Study Investigators: Kidney function and cardiovascular disease in the hypertensive population: the ERIC-HTA study. J Hypertens 2006;24:663–669.

2 Mathew K, Garratt N, Holmes DR, Best PJ, Singh M, Bell MR, Barsness GW, Verghese C, Rihal S, Textor SC, Grill DE, Berger PB, Ting HH: Coronary intervention incidence and prognostic importance of acute renal failure after percutaneous coronary intervention. Circulation 2002;105:2259–2264.

3 Ronco C, McCullough P, Anker SD, Anand I, Aspromonte N: Cardio-renal syndromes: report from the consensus conference of the Acute Dialysis Quality Initiative. Eur Heart J 2010;31:703–771.

4 Go AS, Chertow GM, Dongjie Fan D, McCulloch CE, Hsu CY: Chronic kidney disease and the risks of death, cardiovascular events, and hospitalization. N Engl J Med 2004;351:1296–1305.

5 Beattie JN, Soman SS, Sandberg KR, Yee J, Borzak S, Garg M: Determinants of mortality after myocardial infarction in patients with advanced renal dysfunction. Am J Kidney Dis 2001;37:1191–1200.

6 Anavekar NS, McMurray JJ, Velazquez EJ, Solomon SD, Kober L, Rouleau JL: Relation between renal dysfunction and cardiovascular outcomes after myocardial infarction. N Engl J Med 2004;351:1285–1295.

7 Rostand SG, Kirk KA, Rutsky EA: Dialysis-associated ischemic heart disease: insights from coronary angiography. Kidney Int 1984; 25, 653–659.

8 Joki N, Hase H, Nakamura R: Onset of coronary artery disease prior to initiation of hemodialysis in patients with end-stage renal disease. Nephrol Dial Transplant 1997;12: 718–723.

9 Ohtake T, Kobayashi S, Moriya H: High prevalence of occult coronary artery stenosis in patients with chronic kidney disease at the initiation of renal replacement therapy: an angiographic examination. J Am Soc Nephrol 2005;16:1141–1148.

10 Dries DL, Exner DV, Domanski MJ, Greenberg B, Stevenson LW: The prognostic implications of renal insufficiency in asymptomatic and symptomatic patients with left ventricular systolic dysfunction. J Am Coll Cardiol 2000;35:681–689.

11 Wang LW, Fahim MA, Hayen A, Mitchell RL, Baines L, Lord S, Craig JC, Webster AC: Cardiac testing for coronary artery disease in potential kidney transplant recipients. Cochrane Database Syst Rev 2001;7:1–104.

12 Chonchol M, Whittle J, Desbien: Chronic kidney disease is associated with angiographic coronary artery disease. Am J Nephrol 2008;28:354–360.

13 Charytan DM, Kuntz RE, Garshick M, Candia S, Khan F, Mauri L: Location of acute coronary artery thromboses in patients with and without chronic kidney disease. Kidney Int 2009;75:80–87.

14 Whalley GA, Marwick TH, Doughty RN, Cooper BA, Johnson DW, Pilmore A, Harris DC, Pollock CA, Collins JF: IDEAL effect of early initiation of dialysis on cardiac structure and function: results from the echo substudy of the IDEAL trial. Am J Kidney Dis 2013;61: 262–270.

15 Stary HC, Chandler AB, Dinsmore RE, Fuster V, Glagov S, Insull W, Rosenfeld ME, Schwartz CJ, Wagner WD, Wissler RW: A definition of advanced types of atherosclerotic lesions and a histological classification of atherosclerosis: a report from the Committee on Vascular Lesions of the Council on Arteriosclerosis, American Heart Association. Circulation 1995;92:1355–1374.

16 McCullough PA, Agrawal V, Danielewicz E, Abela GS: Accelerated atherosclerotic calcification and Monckeberg's sclerosis: a continuum of advanced vascular pathology in chronic kidney disease. Clin J Am Soc Nephrol 2008;3:1585–1598.

17 Gross ML, Meyer HP, Ziebart H, Rieger P, Wenzel U, Amann K, Berger I, Adamczak M, Schirmacher P, Ritz E: Calcification of coronary intima and media. Clin J Am Soc Nephrol 2007;2:121–134.

18 Ragosta M, Samady H, Isaacs RB, Lawrence B, Gimple W, Sarembock AJ, Powers ER: Coronary flow reserve abnormalities in patients with diabetes mellitus who have end-stage renal disease and normal epicardial coronary arteries. Am Heart J 2004;147: 1017–1023.
19 Garland JS, Holden RM, Groome PA, Lam M, Nolan RL, Morton R, Pickett W: Prevalence and associations of coronary artery calcification in patients with stages 3–5 CKD without cardiovascular disease. Am J Kidney Dis 2008;52:849–858.
20 Adragao T, Pires A, Branco P, Castro R, Oliveira A, Nogueira C, Bordalo J, Curto JD, Prata MM: Ankle-brachial index, vascular calcifications and mortality in dialysis patients. Nephrol Dial Transplant 2012;27: 318–325.
21 Moe SM, O'Neill KD, Duan D, Ahmed S, Chen NX, Leapman SB, Fineberg N, Kopecky K: Medial artery calcification in ESRD patients is associated with deposition of bone matrix proteins. Kidney Int 2002;61:638–647.
22 Campean V, Nonnast-Daniel B, Garlichs C, Gross ML, Amann K: CD40–CD154 expression in calcified and non-calcified coronary lesions of patients with chronic renal failure. Atherosclerosis 2007;190:156–166.
23 McGregor E, Jardine AG, Murray LS, Dargie HJ, Rodger RS, Junor BJ: Preoperative echocardiographic abnormalities and adverse outcome following renal transplantation. Nephrol Dial Transplant 1998;13:1499–1505.
24 London GM, Pannier B, Guerin AP, Blacher J, Marchais SJ, Darne B: Alterations of left ventricular hypertrophy in and survival of patients receiving hemodialysis: follow-up of an interventional study. J Am Soc Nephrol 2001;12:2759–2767.
25 Laurent S, Cockcroft J, Van Bortel L, Boutouyrie P, Giannattasio C, Hayoz D, Pannier B, Vlachopoulos C, Wilkinson I, Struijker-Boudier H: Expert consensus document on arterial stiffness: methodological issues and clinical applications. Eur Heart J 2006;27:2588–2605.
26 Foley RN, Parfey PS, Harnett JD, Kent GM, Martin GJ, Murray DC, Barre PE: Clinical and echocardiographic disease in patients starting end-stage renal disease therapy. Kidney Int 1995;47:186–192.
27 Foley RN, Curtis BM, Randell EW, Parfey PS: Left ventricular hypertrophy in new hemodialysis patients without symptomatic cardiac disease. Clin J Am Soc Nephrol 2010; 5:805–813.
28 Lekawanvijit S, Adrahtas A, Kelly DJ, Kompa AR, Wang BH, Krum H: Does indoxyl sulfate, a uremic toxin, have direct effects on cardiac fibroblasts and myocytes? Eur Heart J 2010;31:1771–1779.
29 McIntyre CW, Burton JO, Selby NM, Leccisotti L, Korsheed S, Baker CSR, Camici PG: Hemodialysis-induced cardiac dysfunction is associated with an acute reduction in global and segmental myocardial blood flow. Clin J Am Soc Nephrol 2008;3: 19–26.
30 Parfey PS, Harnett JD, Foley RN: Heart failure and ischemic heart disease in chronic uremia. Curr Opin Nephrol Hypertens 1995; 4:105–110.
31 Blacher J, Safar ME, Guerin AP, Pannier B, Marchais SJ, London GM: Aortic pulse wave velocity index and mortality in end-stage renal disease. Kidney Int 2003;63:852–1860.
32 Bogrov AY, Shapiro JI: Endogenous digitalis: pathophysiologic roles and therapeutic applications. Nat Clin Pract Nephrol 2008; 4378–4392.
33 Wali RK, Wang GS, Gottlieb SS, et al: Effect of kidney transplantation on left ventricular systolic dysfunction and congestive heart failure in patients with end-stage renal disease. J Am Coll Cardiol 2005;45:1051–1060.
34 Amann K, Breitbach M, Ritz E, Mall G: Myocyte/capillary mismatch in the heart of uremic patients. J Am Soc Nephrol 1998;9: 1018–1022.
35 Raine AEG, Seymour AML, Roberts AFC, Radda GK, Ledingham JGG: Impairment of cardiac function and energetics in experimental renal failure. J Clin Invest 1993; 92:2934–2940.
36 Herzog CA, Asinger RW, Berger AK, Charytan DM, Díez J, Hart RG, Eckardt KU, Kasiske BL, McCullough PA, Passman RS, DeLoach SS, Pun PH, Ritz E: Cardiovascular disease in chronic kidney disease. A clinical update from Kidney Disease: Improving Global Outcomes (KDIGO). Kidney Int 2011; 80:572–586.

37 Mall G, Huther W, Schneider J, Lundin P, Ritz E: Diffuse intermyocardiocytic fibrosis in uraemic patients. Nephrol Dial Transplant 1990;5:39–44.
38 Vanholder R, De Smet R, Glorieux G, Argiles A, Baurmeister U, Brunet P, Clark W, Cohen G, De Deyn PP, Deppisch R, Descamps-Latscha B, Henle T, Jorres A, Lemke HD, Massy ZA, Passlick-Deetjen J, Rodriguez M, Stegmayr B, Stenvinkel P, Tetta C, Wanner C, Zidek W: Review on uremic toxins: classification, concentration, and inter individual variability. Kidney Int 2003;63:1934–1943.
39 Meert N, Schepers E, De Smet R, Argiles A, Cohen G, Deppisch R, Drueke T, Massy Z, Spasovski G, Stegmayr B, Zidek W, Jankowski J, Vanholder R: Inconsistency of reported uremic toxin concentrations. Artif Organs 2007;31:600–611.
40 Miyazaki T, Ise M, Seo H, Niwa T: Indoxyl sulfate increases the gene expressions of TGF-β_1, TIMP-1 and pro-α1(I) collagen in uremic rat kidneys. Kidney Int Suppl 1997; 62:S15–S22.
41 Milting H, Ellinghaus P, Seewald M: Plasma biomarkers of myocardial fibrosis and remodeling in terminal heart failure patients supported by mechanical circulatory support devices. J Heart Lung Transplant 2008;27: 589–596.
42 Sharma UC, Pokharel S, van Brakel TJ: Galectin-3 marks activated macrophages in failure-prone hypertrophied hearts and contribute to cardiac dysfunction. Circulation 2004;110:3121–3128.
43 Lok DJ, Van Der Meer P, de la Porte PW, Lipsic E, Van Wijngaarden J, Hillege HL, van Veldhuisen DJ: Prognostic value of galectin-3, a novel marker of fibrosis, in patients with chronic heart failure: data from the DEAL-HF study. Clin Res Cardiol 2010; 99:323–328.
44 Lin YH, Lin LY, Wu YW, Chien KL, Lee CM, Hsu RB, Chao CL, Wang SS, Hsein YC, Liao LC, Ho YL, Chen MF: The relationship between serum galectin-3 and serum markers of cardiac extracellular matrix turnover in heart failure patients. Clin Chim Acta 2009;409: 96–99.
45 Gopal DM, Kommineni M, Ayalon N, Koelbl C, Ayalon R, Biolo A, Dember LM, Downing J, Siwik DA, Liang CS, Colucci WS: Relationship of plasma galectin-3 to renal function in patients with heart failure: effects of clinical status, pathophysiology of heart failure, and presence or absence of heart failure. J Am Heart Assoc 2012;1:1–7.
46 Faul C, Amaral AP, Oskouei B, Hu MC, Sloan A, Isakova T, Gutiérrez OM, Aguillon-Prada R, Lincoln J, Hare JM, Mundel P, Morales A, Scialla J, Fischer M, Soliman EZ, Chen J, Go AS, Rosas SE, Nessel L, Townsend RR, Feldman HI, St John Sutton M, Ojo A, Gadegbeku C, Di Marco GS, Reuter S, Kentrup D, Tiemann K, Brand M, Hill JA, Moe OW, Kuro-O M, Kusek JW, Keane MG, Wolf M: FGF23 induces left ventricular hypertrophy. J Clin Invest 2011;121:4393–4408.
47 Gutiérrez OM, Isakova T, Rhee E, Shah A, Holmes J, Collerone G, Jüppner H, Wolf M: Fibroblast growth factor-23 mitigates hyperphosphatemia but accentuates calcitriol deficiency in chronic kidney disease. J Am Soc Nephrol 2005;16:2205–2215.
48 Shimada T: Circulating fibroblast growth factor 23 in patients with end-stage renal disease treated by peritoneal dialysis is intact and biologically active. J Clin Endocrinol Metab 2010;95:578–585.
49 Gutiérrez OM, Januzzi JL, Isakova T, Laliberte K, Smith K, Collerone G, Sarwar AS, Hoffmann U, Coglianese E, Christenson R, Wang TJ, deFilippi C, Wolf M: Fibroblast growth factor 23 and left ventricular hypertrophy in chronic kidney disease. Circulation 2009;119:2545–2552.

James A. Tumlin, MD
Internal Medicine Department, Division of Clinical Research
University of Tennessee College Medicine
Chattanooga, TN 37408 (USA)
E-Mail jamestumlinmd@nephassociates.com

McCullough PA, Kellum JA, Mehta RL, Murray PT, Ronco C (eds): ADQI Consensus on AKI Biomarkers and Cardiorenal Syndromes. Contrib Nephrol. Basel, Karger, 2013, vol 182, pp 174–194
DOI: 10.1159/000349970

Cardiorenal Syndrome Type 5: Clinical Presentation, Pathophysiology and Management Strategies from the Eleventh Consensus Conference of the Acute Dialysis Quality Initiative (ADQI)

Ravindra L. Mehta[a] · Hamid Rabb[b] · Andrew D. Shaw[c] · Kai Singbartl[d] · Claudio Ronco[e] · Peter A. McCullough[f] · John A. Kellum[g] · for the Acute Dialysis Quality Initiative (ADQI) 11 Working Group

[a]Department of Medicine, University of California San Diego, San Diego, Calif., USA; [b]Department of Medicine, Johns Hopkins University, Baltimore, Md., USA; [c]Department of Anesthesiology, Duke University Medical Center, Durham, N.C., USA; [d]Department of Anesthesiology, Penn State College of Medicine, Milton S. Hershey Medical Center, Hershey, Pa., USA; [e]Department of Nephrology, International Renal Research Institute (IRRIV), San Bortolo Hospital, Vicenza, Italy; [f]St. John Providence Health System, Warren, Mich., Providence Hospitals and Medical Centers, Southfield and Novi, Mich., St. John Macomb Oakland Center, Madison Heights, Mich., St. John Hospital and Medical Center, Detroit, Mich., USA; [g]Clinical Research, Investigation, and Systems Modeling of Acute Illness (CRISMA) Center, Department of Critical Care Medicine, University of Pittsburgh, Pittsburgh, Pa., USA

Abstract

Cardiorenal syndrome type 5 (CRS-5) includes conditions where there is a simultaneous involvement of the heart and kidney from a systemic disorder. Given the wide spectrum of diseases that contribute to CRS-5, several pathophysiological mechanisms are invoked representing the response of the heart and kidney to the contributing disorder that is ongoing. The nature, duration and the underlying condition of the heart and kidney strongly influence the clinical course of CRS-5. In this paper we discuss the pathophysiology of CRS-5 in the setting of sepsis as a model system for CRS-5 providing a brief overview of strategies for monitoring and therapeutic intervention. We offer a framework for reference for considering other disorders leading to CRS-5 where the development of cardiac and renal dysfunction is more insidious.

ADQI 11 Workgroup members are listed in Appendix 2.

Type 5 cardiorenal syndrome (CRS-5) occurs when an overwhelming insult leads to the simultaneous development of acute kidney injury (AKI) and acute cardiac dysfunction. In the literature, CRS-5 most commonly encompasses a wide spectrum of disorders that acutely involve the heart and kidney such as sepsis and drug toxicity where both the heart and the kidney are involved secondarily to the underlying process [1]. CRS-5 may develop in the setting of previously impaired organ function or when there is no discernible evidence of prior abnormality. The sequence of organ involvement can vary depending on the acuity and nature of the underlying disorder. The time sequence for developing CRS-5 depends on the underlying disease and is influenced by the underlying level of cardiac and renal function.

Given the multitude of contributing factors and the time sequence of events in CRS-5, it is challenging to identify the underlying pathophysiological mechanisms and develop a strategy for diagnostic and therapeutic intervention. In this paper we provide a framework for considering the pathophysiology of CRS-5 using severe sepsis and septic shock as model disorders in which we discuss the relevant systems pathophysiology, molecular pathways involved diagnostic approach and therapeutic considerations.

Methods

We performed a systemic review of the literature pre-conference, as described elsewhere. Specifically, we used the terms 'cardiorenal syndrome type 5' and 'acute kidney injury', combined with 'sepsis', 'liver failure', 'lupus', 'pathophysiology' and 'mechanisms'. In view of the volume of the retrieved literature, only representative publications are cited in this review. Furthermore, we opted to focus this summary to the clinical scenario of acute CRS-5 using sepsis as a model system.

Results

Based on the literature identified prior to the conference, the following key questions were considered:

1. What is the underlying pathophysiology of CRS-5?
    a. What are the predisposing and modifying factors?
    b. What are the direct effects of the underlying disorder on the heart and kidney?
    c. Does involvement of one organ contribute to dysfunction in the other?
    d. Which cellular and molecular mechanisms are involved?

e. What is the influence of process of care on pathophysiology, e.g. fluid resuscitation?
2. What are the best techniques for the diagnosis and monitoring of CRS-5?
   a. To evaluate the nature and severity of cardiac and renal damage.
   b. To assess progression and resolution of organ dysfunction.
3. What are the targets and best strategies for therapeutic intervention?
   a. Management of underlying disease.
   b. Treatment of cardiac and renal dysfunction.

Q1. What Is the Underlying Pathophysiology of Cardiorenal Syndrome Type 5?

a. What Are the Predisposing and Modifying Factors?

The pathophysiology of CRS-5 depends on the underlying disease state, the time frame for development and the context in which it is encountered. Acute CRS-5 results from systemic processes, e.g. sepsis, infections, drugs, toxins and connective tissue disorders such as lupus, Wegener's granulomatosis, and sarcoidosis. While there is some overlap in these conditions, to a large extent, the time frame for development and resolution represent the underlying pathophysiological mechanisms. For instance, in sepsis-induced acute CRS-5 there is a fulminant disease process with dramatic impact on both kidney and heart, with obvious clinical manifestations. In contrast, in cirrhotic liver disease, CRS-5 has a more insidious onset and the kidney and cardiac dysfunction may develop slowly until a 'tipping point' is reached and full-blown decompensation occurs. Several factors influence the course of CRS-5 (table 1). The inducing events and the immunological and physiological responses are conditioned by the underlying condition of the heart and kidney. Although the mechanisms invoked in acute and chronic CRS-5 are somewhat different (fig. 1, 2), the nature, severity and duration of organ dysfunction are also influenced by the process of care to manage the condition (table 1).

In order to understand the underlying pathophysiology and identify opportunities for intervention it is helpful to consider the temporal sequence of CRS-5 as characterized by the natural history and management of the underlying disorder (table 2). Acute CRS-5 develops along a temporal framework consisting of hyperacute (0–72 h after diagnosis), acute (3–7 days), subacute (7–30 days) and chronic (30+ days) phases (fig. 3). Most published human literature refers to the hyperacute phase, since this has been the focus of most of the interventional clinical trials conducted to date in sepsis. Chronic CRS-5, e.g. in cirrhotic patients, similarly demonstrates distinct phases of pre-ascites, diuret-

Table 1. Predisposing and modifying factors influencing development of sepsis induced CRS-5

Patient attributes	Process of care
Inducing event and immunological response	Resuscitation – Volume effect on tissue edema and renal venous hypertension – Abdominal compartment syndrome – Glycocalyx
Underlying condition of heart and kidney	Search for primary focus – Contrast use
Physiological responses – Peripheral vasodilation – Compensatory hormonal response – Increased vascular permeability – Mitochondrial dysfunction and tissue hypoxia – Cardiac filling – Renal hypoperfusion	Drug dosing – Antibiotics – Sedatives – Vasopressors and ionotropes
	General supportive care – Fluids – Inotropes, pressors
	Specific interventions – Ventilator – Surgery – RRT

ic responsive and diuretic refractory ascites prior to the development of hepatorenal syndrome although the time sequence is quite variable (fig. 4). In most cases of chronic CRS-5 there is usually a precipitating event that brings the condition to attention. For instance in cirrhosis this is often an infection such as spontaneous bacterial peritonitis. Thus an acute CRS-5 can be superimposed on a chronic indolent process. Although we recognize the importance of all phases of CRS-5, we focus here on the pathophysiology of acute CRS-5 because of its immediate relevance for intensivists, nephrologists and cardiologists.

b. What Are the Direct Effects of Underlying Disorder on the Heart and Kidney?

With respect to the cardiorenal syndrome, changes in systems physiology during sepsis can result from the systemic effects of sepsis itself, from the septic injury to 'systemic pathways', or from the organ cross-talk between injured heart and kidney.

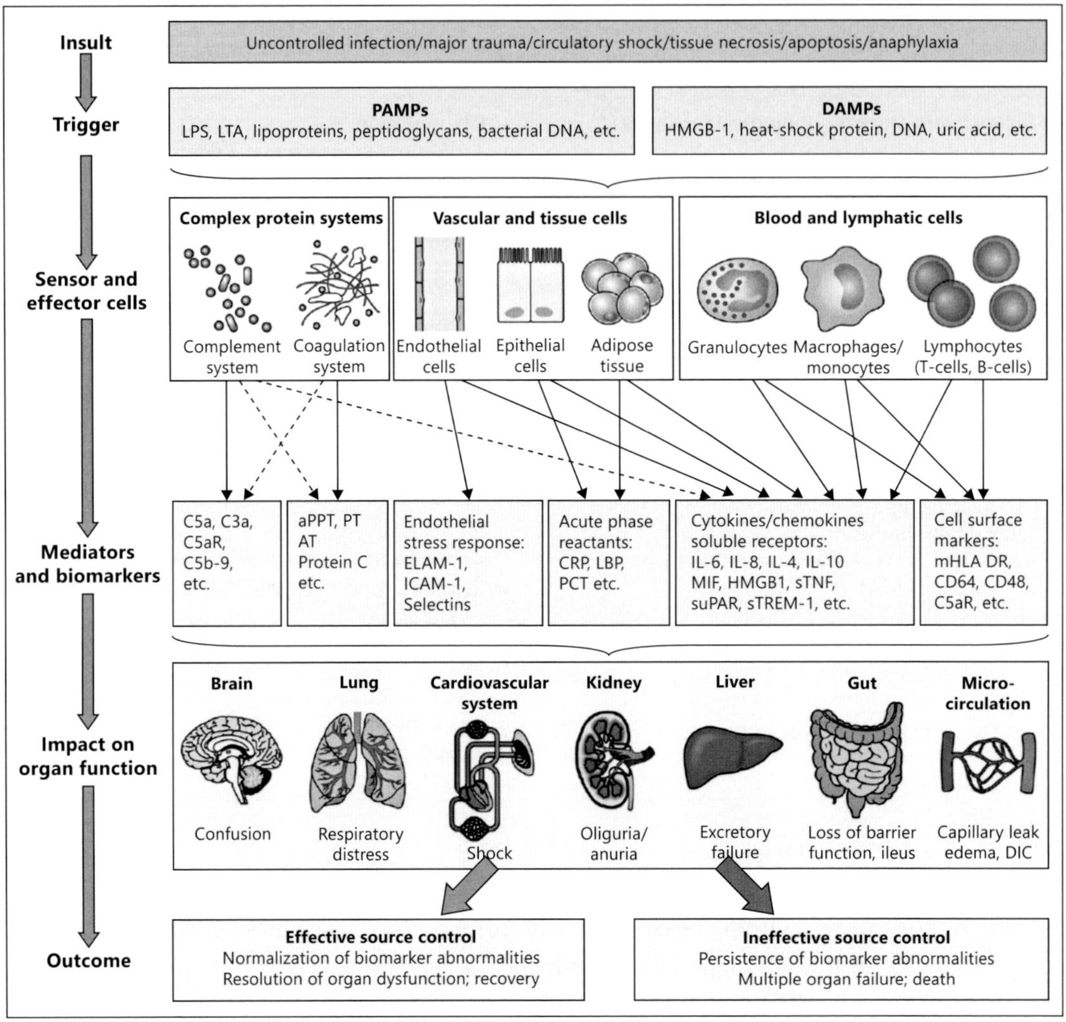

Fig. 1. Pathophysiology of sepsis-induced organ dysfunction. From [49] with permission.

An overall key feature of sepsis is the dissociation between the systemic circulation and the microcirculation in various organs [2]. Especially in the early phases of sepsis, profound microcirculatory changes can develop, despite apparently normal systemic hemodynamics [1–3]. Microcirculatory changes, such as lower blood flow velocities and heterogeneous perfusion patterns, strongly correlate with morbidity and mortality rates [4].

Septic Injury of the Heart (Septic Cardiomyopathy). Septic cardiomyopathy (SCM) represents one of the major predictors of morbidity and mortality of sepsis and is present in nearly 50% of all patients [5, 6]. Sepsis can cause both the

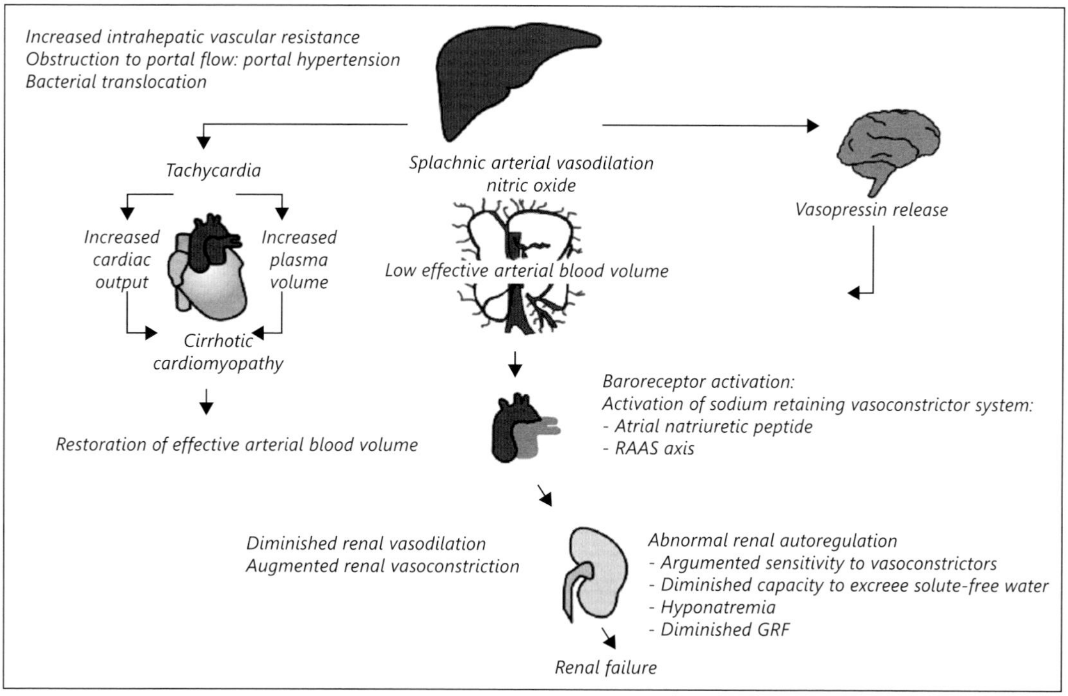

Fig. 2. Pathophysiology of cirrhosis-induced CRS-5 [from 91, with permission].

Table 2. Temporal considerations in pathophysiology of CRS-5

Attribute	CRS-5 acute (sepsis) (fig. 1)	CRS-5 chronic (cirrhosis) (fig. 2)
Time for organ dysfunction	Short: hours to days	Long: weeks to months
Underlying organ function	May be superimposed on underlying cardiac and kidney disease	Heart and kidney have adaptive responses that fail over time
Sequence of organ involvement	Generally simultaneous or in close proximity to each other	One organ precedes the other, e.g. cardiac dysfunction precedes renal in cirrhosis
Underlying disease	Systemic event contributes to CRS5	Precipitating events can transition to an acute deterioration in CRS5, e.g. GI bleeding can precipitate hepatorenal syndrome
Pathophysiology	Direct effects on organs	Failure of adaptive responses over time
Mechanisms	Determined by underlying disease	Determined by adaptive changes
Reversibility	Possible with control of sepsis and organ support	Limited unless there is replacement of diseased organ, e.g. liver transplant

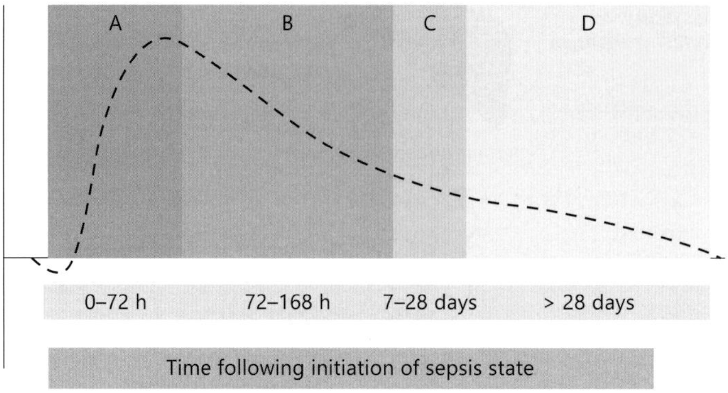

	Hyperacute	Acute	Subacute	Chronic
Systemic hemodynamics	Normal-↑↑↑*	Normal-↑ or ↓*	Normal*	Normal
Microcirculation	Maldistribution#	???	???	???
Autonomic nervous system	↓ HRV#*	Normal-↓ HRV#*	???	???
RAAS	↑↑↑#*	Normal-↑#*	???	???
HPA	↓↓↓*	↓↓-↓*	???	???

*Clinical evidence
#Experimental evidence

Fig. 3. Pathophysiologic changes in sepsis-induced acute CRS-5. Organ dysfunction in sepsis can be considered in several different phases that reflect the severity of the disease. The underlying pathophysiological mechanisms are different in each phase and thus provide specific opportunities for targeted diagnostic and therapeutic strategies. Phase A: 0–72 h = hyperacute phase; phase B: 72–168 h = acute phase; phase C: 7–28 days = subacute phase; phase D: >28 days = chronic. RAAS = Renin-angiotensin-aldosterone system; HPA = Hypothalmus-pituitary gland-adrenal gland axis; HRV = Heart rate variation.

left and the right ventricle to dilate and develop decreased ejection fractions, leading to impaired responsiveness to fluid and catecholamine treatment [7]. Although SCM can become so severe mimicking cardiogenic shock, it is reversible within 7–10 days in most cases [5, 6]. Moreover, as long as hypovolemia is corrected, tachycardia and reduced vascular tone allow for a maintained or even increased cardiac output in many patients. Contrary to early wisdom, myocardial blood flow or energy metabolism do not seem to play a role in SCM [8]. On the other side, myocardial depressants, including pro-inflammatory cytokines and complement factors, have emerged as crucial factors in the development of SCM [9–11].

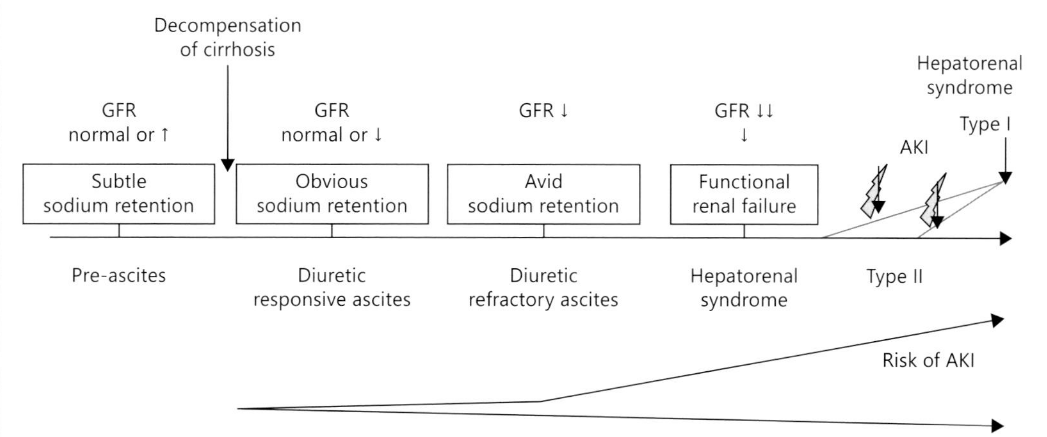

	Compensated cirrhosis	Diuretic-responsive ascites	Diuretic-resistant ascites	HRS
Splanchnic/systemic arterial vasodilation	Normal/+	+	++	+++
Effective circulating volume	Normal	–	– –	– – –
Renin, aldosterone, vasopressin, norepinephrine	Normal	+	++	+++
Renal sodium retention/plasma volume	+	++	+++	++++
Renal vasoconstriction	Normal	Normal	++	++++

Fig. 4. Chronic CRS-5 hemodynamic changes during different stages of cirrhosis. From [92] with permission.

Septic Kidney Injury. In addition to cellular and molecular changes in the kidney, which are discussed below, there are also alterations to the intra-renal blood flow during sepsis. These changes are uncoupled from the systemic hemodynamic changes occurring during sepsis [12–14]. The exact nature of these changes, however, remains highly controversial. A lack of sound clinical data as well as the heterogeneous quality and limited applicability of various animal models do not allow us to develop a clear concept. A recent experimental study comparing two different sepsis models in pigs might be the first step to overcome this problem [12]. Both experimental design and findings resemble the heterogeneity seen in septic patients very well. Irrespective of systemic hemodynamics (normo- or hyperdynamic), only animals that developed septic AKI also demonstrated an increase in renal vascular resistance. Moreover, only animals with septic AKI also exhibited early rises in pro-inflammatory cytokines, e.g. IL-6, and in markers of oxidative stress. The excellent agreement of these findings with those from observational clinical studies [15, 16] allows us to formulate strong hypotheses that require further testing in larger clinical studies.

Neurogenic and Endocrine Alterations. Sepsis also affects central structures or pathways, including the autonomic nervous system (ANS), the renin-angiotensin-aldosterone system (RAAS) and the hypothalamus-pituitary gland-adrenal gland axis (HPA), which in turn impact cardiac and/or renal function. Sepsis causes ANS dysfunction [17–19], the severity of which correlates with morbidity and mortality. A hallmark of autonomic dysfunction during sepsis is the decreased heart rate variability. A decrease or even loss of heart rate variability (HRV) during sepsis is associated with the release of inflammatory mediators, e.g. IL-6, IL-10 and CRP. Data with respect to kidney-related changes in ANS during sepsis is limited to animal studies. Here, sepsis-induced changes in renal sympathetic nerve activity did not seem to affect renal blood flow [20].

Sepsis activates the RAAS, reflecting the body's attempt to restore and maintain a sufficient blood pressure. Although counterintuitive, recent experimental and limited clinical data suggest that blockade of the RAAS might be beneficial, as RAAS activation has also been implicated in endothelial dysfunction, organ failure and even mortality during severe sepsis [21–24]. Experimental studies further suggest deleterious effects of RAAS activation on renal function during sepsis [13, 25–27]. The administration of ACE inhibitors improves creatinine clearance and urine output during experimental bacteremia; the application of selective angiotensin II type 1 receptor antagonist improves renal blood flow and oxygenation during experimental endotoxemia.

Sepsis causes complex alterations of HPA and glucocorticoid signaling, leading to severe adrenal insufficiency in some patients [28]. Adrenal insufficiency in turn gives rise to increased production of pro-inflammatory cytokines, free radicals and prostaglandins as well as inhibition of chemotaxis and expression of adhesion molecules. Administration of moderate-dose glucocorticoids for 7 days can reduce the need for vasopressors and intensive care unit (ICU) length of stay [29, 30].

There is no sound data regarding the effects of adrenal insufficiency on renal function.

c. Does Involvement of One Organ Contribute to Dysfunction in the Other?

Among the most provocative yet complex questions in multiorgan failure is to determine the precise role of the failing heart or kidney in dysfunction of either organ. The plethora of effects of sepsis on the function of various organs, including heart and kidneys, makes it very difficult to differentiate between the effects of sepsis itself and the effects of inter-organ cross-talk. However, some effects of inter-organ cross-talk can be postulated from the sepsis-specific pathophysiology: (i) a reduced cardiac output will lead to reduced renal perfusion, further aggravating sepsis-induced kidney injury; (ii) fluid overload due to AKI can lead to CHF in already dilated and hypocontractile heart, and (iii) AKI-induced metabolic acidosis can impair contractility and increases the heart rate, worsening myocardial workload.

Limited experimental data allow us to hypothesize that AKI exerts cardio-depressing effects via remote induction of pro-inflammatory and pro-apoptotic pathways [31].

Beyond the hemodynamic effects of the failing heart on renal circulation, there are cardiac changes due to impaired fluid clearance by the kidney. However, AKI itself is well established experimentally to lead to distant organ function [32]. In a murine model, AKI led to decreased cardiac contractility and apoptosis, which was partially abrogated by anti-TNF treatment [31]. AKI also led to cardiac hypertrophy [33]. AKI can cause an increase in cardiac macrophages [34]. Better studied has been the effect of AKI on lung function, where there is an increase in microvascular permeability, endothelial cell dysfunction, and caspase-mediated apoptosis [35]. A well-orchestrated molecular response occurs in lung independent of volume overload [36], and this signature has been localized to endothelial cell [37]. The brain is particularly affected by AKI, and this may in turn affect systemic neuroendocrine response in sepsis [38]. Though this is a particularly challenging area of research, it is potentially among the most important as this could decrease the severe mortality associated with combined acute heart and kidney failure by guiding extracorporeal support devices which cleanse circulating factors that enable deleterious inter-organ cross-talk.

d. Which Cellular and Molecular Mechanisms Are Involved?

During combined acute cardiac and kidney dysfunction, such as in sepsis, there are marked cellular and molecular changes in each organ. There is an important time-specific pattern for these changes, but this aspect has not been sufficiently studied. Furthermore, it is difficult, except using highly controlled experimental models, to tease out the role of the primary infection versus the role of each failing organ on the other.

Activation and induction of cytokines, leukocytes, toll receptors is well established in heart and kidney during sepsis [39]. Notable are pro-inflammatory cytokines TNF and IL-6, while IL-10 has been implicated as anti-inflammatory. Macrophages, neutrophils, lymphocytes, and more recently T regs, have been implicated. There is an increasing body of evidence supporting the role of danger and pattern-associated receptors, DAMPS and PAMPs, and Toll receptors are among the most important pathophysiologically (fig. 1) [40]. Abnormalities in oxidative stress are likely key mechanisms as well. These range from mitochondrial dysfunction [41]), to alteration in antioxidant stress enzymes, particularly those mediated by the Nrf2-Keap pathway [42].

Given the fundamental contractile function of the heart, there are specific molecular pathways that are deranged. Cardiac muscle protein expression, particularly actin and myosin are abnormal in sepsis. Membrane-associated proteins, especially dystrophin, regulate cell shape, mechanical strength and cardiomyocyte transduction of contractile force. The mean amounts of dystrophin and associated glycoproteins in significant reduced in septic hearts [39]. Intercalated discs are the sites where cardiomyocytes connect to each other and both ensure mechanical coupling plus enable electrical signal propagation. Components of intercalated discs including connexin-43 and N-cadherin are markedly reduced in sepsis, with dehiscence of the junctions during sepsis. L-type calcium currents, important in cardiac function, are reduced during septic shock [43].

In the kidney, geared towards filtration, fluid and electrolyte regulation, there are specific tubular changes after sepsis. Lipopolysaccharide directly alters HCO_3^- transport through toll receptors, leads to abnormalities in acidification of urine [44]. Sepsis leads to a decline in autophagy in the proximal tubule, which in turn contributes to proximal tubule function [45]. LPS also modifies the important glomerular protein megalin, which can contribute to an increase in urine albumin [46].

e. What Is the Influence of Process of Care on Pathophysiology?

- The management of sepsis can contribute to the development of CRS-5. Fluid resuscitation coupled with the increased vascular permeability can lead to organ and tissue edema and result in an abdominal compartment syndrome with an increase in renal venous pressure and reduced renal perfusion [47]. The utilization of contrast agents for imaging studies can similarly lead to myocardial depression and tubular toxicity. There is a prolonged half life of drugs due to a reduction in hepatic and renal clearance that can result in organ toxicity. All these factors can influence the development and course of CRS-5.

Q2. What Are the Best Techniques for the Diagnosis and Monitoring of CRS-5?

In considering a diagnostic approach to CRS-5 it is important to decide what the clinician is trying to achieve at each stage of the disease process. Initially the emphasis is on diagnosis (of severe sepsis and septic shock), organ function assessment, risk prediction and treatment guidance. Attention then turns to monitoring organ function (using both physiological and biochemical tools) while the patient is stabilized and further organ dysfunction prevented where possible. Assuming the patient begins to recover, the emphasis turns to assessment of residual organ function impairment and assessment of the long-term impact of the disease on the patient's life. Here we discuss these in the context of CRS-5.

a. To Evaluate the Nature and Severity of Cardiac and Renal Damage

Diagnosis of Severe Sepsis and Septic Shock. For practical purposes, when systemic inflammation occurs in the setting of known or suspected infection, the diagnosis of sepsis is made. Once organ dysfunction develops the diagnosis progresses to severe sepsis, and when the systemic blood pressure cannot be maintained without vasoactive therapy then septic shock is present. Systemic inflammation is usually said to be present when two or more of the systemic inflammatory response syndrome (SIRS) criteria are present: body temperature <36°C (96.8°F) or >38°C (100.4°F); heart rate >90 beats/min; tachypnea (>20 breaths/min) or arterial partial pressure of carbon dioxide <32 mm Hg; white blood cell count $<4 \times 10^9$ cells/l or $>12 \times 10^9$ cells/l; or the presence of >10% immature neutrophils (band forms) although many sepsis trials have required three of these criteria to be present in order to increase the specificity of the diagnosis [48]. Reinhart et al. [49] present an excellent review of the current state of the art regarding biomarkers for the diagnosis and management of sepsis – readers are referred to this for detailed information regarding the place of candidate markers such as lipopolysaccharide binding protein, procalcitonin, acute phase reactants, cytokines, chemokines and '-omic' techniques including the use of microarrays for pathogen determination and treatment.

Assessment of Cardiac and Renal Function. Assessment and monitoring of cardiac dysfunction in CRS-5 relies on the same techniques as in other causes of acute myocardial dysfunction. Biochemical assays based on natriuretic peptides and troponins provide evidence of cardiac chamber morphological change and myocyte leak respectively. Traditional markers such as leukocytosis and C-reactive protein are non-specific for myocardial assessment, and most clinicians rely on functional monitoring to guide assessment and management. The cardiomyopathy of sepsis is complex and incompletely understood [50], but it is fair to say that in the early (unresuscitated) stages there is generalized depression of cardiac function, manifesting clinically as a low cardiac output state. Following fluid therapy the situation often changes to a picture more typical of distributive shock – rapid bounding pulses, increased cardiac output and systemic vasodilation. Most ICUs no longer use pulmonary artery catheters to guide diagnosis and hemodynamic management in septic shock [51], and thus this is almost certainly the case for CRS-5 too. Lastly, there is a growing appreciation of the value of echocardiography in the assessment and management of ICU patients with acute myocardial dysfunction [52]. Bedside assessment of myocardial filling status and systolic function is particularly helpful even if a full study is not obtainable.

Assessment of Renal Function. Diagnosis and assessment of renal function in sepsis-induced CRS-5 is the same as in other causes of AKI. The clinical mainstay is still an acute change in serum creatinine levels, and consensus has been reached with the RIFLE [53], AKIN [54] and KDIGO criteria [55] essentially describing similar acute reductions in renal function. These systems also provide a means of assessing severity and have served to move the clinical research agenda forwards from recognition of the disease to risk assessment, assessment of treatment effect and transition to clinically important endpoints. In CRS-5, several novel biomarkers of AKI have been reported to have clinical utility. For reviews of these see Comnick and Ishani [56] and Cruz et al. [57]. Several new markers of renal function and AKI risk prediction are now in clinical use in Europe and time will tell if they have true clinical utility. Cystatin C is still the only new biomarker approved for diagnostic assessment in the USA at the time of writing. For the time being at least therefore it is likely that RIFLE, AKIN and KDIGO criteria using serum creatinine and urine output will continue to be used for diagnosis and monitoring of AKI in CRS-5.

b. To Assess Progression and Resolution of Organ Dysfunction

Risk Prediction. Once the diagnosis of septic shock and CRS-5 is made, organ function has been assessed and the patient is (hopefully) recovering, attention turns to risk prediction and protection of further deteriorations in organ function. In AKI, there are new data suggesting that biomarkers of cell cycle arrest may be able to predict which patients will progress to severe AKI within a few days [93]. These data will need to be confirmed in clinical practice, but the idea that a simple blood test may help clinicians identify those at increased risk of severe disease is certainly attractive. For cardiac risk prognostication, septic shock survivors have been shown to have lower ejection fractions and higher end-diastolic volumes when compared with patients who died, which appears to suggest a possible protective effect of myocardial depression [58]. Rudiger and Singer [50] provide a good discussion of the mechanisms underlying cardiac dysfunction in sepsis.

Table 3. Therapeutic interventions in sepsis-induced CRS-5

Disease modification
Removal of inflammatory mediators Rationale based on targeting removal of inflammatory molecules as molecular weight of most cytokines (e.g. TNF) ranges from 17,000–50,000 Cytokine removal is variable and depends on operational characteristics Removal can be enhanced by – Increasing permeability of membrane (high cut-off membranes, plasma filtration) – Increasing convection (HVHF, pulse HVHF) – Utilizing adsorption (PMMA CHDF, polymyxin, Osiris membrane)
Immunomodulation Rationale based on targeting cellular elements of disease response – Immunoparalysis – Apoptosis – Neutrophil activation Strategies include – Endotoxin removal by selective binding to polymyxin – Anticoagulant-based strategies (citrate enabled selective cytophoretic device) – Plasma filtration coupled to binding to sorbents
Organ support
Cardiac Cardiac dysfunction contributes to shock and poor tissue perfusion Increased vascular permeability results in maldistribution of fluid in various compartments Strategies include – Ionotropes – Vasopressors – Vasodilators
Renal Renal dysfunction contributes to organ failure – Solute clearance – Fluid management – Acid base and electrolyte homeostasis Strategies include – Diuretics – RRT

Q3. What Are the Targets and Best Strategies for Therapeutic Intervention?

Therapeutic interventions in CRS-5 can be focused on disease modification, management of cardiac and renal dysfunction and their consequences (table 3). Decision for interventions should be based on knowledge of the phase of disease, the underlying pathophysiology and anticipated course.

a. Management of Underlying Disease

General Principles. As shown in figure 3, patients in the hyperacute phase will have an emphasis on resuscitation and stabilization of hemodynamic factors coupled with source control. Several publications, e.g. the 'surviving sepsis' guidelines, provide recommendations for the initial management of patients with sepsis and septic shock [51]. Maintaining hemodynamic stability and tissue perfusion are key components for preventing CRS-5 in the hyperacute phase. However, source control and adequate antibiotic coverage need to be addressed. As discussed earlier, often fluid resuscitation results in fluid accumulation and continued maintenance fluids can contribute to fluid overload with deleterious consequences [59–62]. Thus avoiding iatrogenic complications from resuscitation and recognition of failing organ function are critical to reduce the development of CRS-5.

Disease Modification. The pathophysiology and molecular events of sepsis offers several therapeutic targets [63]. As shown in figure 1 the systemic inflammatory response with PAMPS and DAMPS sets into motion the release of several circulating cytokines [49]. Physical removal of cytokines and immunomodulation are the two main approaches that have been tried and are currently being assessed for disease modification based on extracorporeal techniques (table 3). Removal is based on the rationale that cytokine burden can be reduced by utilizing convection [64], high-volume hemofiltration [65, 66], high-permeability membranes [67, 68] or adsorption [69]. High-volume hemofiltration has not been very successful [66, 70], however limited studies have shown benefit with high-permeability membranes and adsorption [71]. An alternate approach is to target cellular elements of the disease response including immunoparalysis, apoptosis and neutrophil activation through endotoxin removal with a polymyxin filter [72] or a citrate anticoagulant-based selective cytopheretic device [73]. Both these techniques have been successful in small pilot studies and pivotal trials are underway. Plasma filtration coupled with sorbents has been studied in experimental models and clinical studies are being initiated [74, 75].

c. Treatment of Cardiac and Renal Dysfunction

Cardiac Support. Managing cardiac dysfunction in the hyperacute and acute phases requires a multipronged approach for maintaining adequate filling pressures with fluid resuscitation and the judicious uses of vasopressors, vasodilators and ionotropes. Several studies have evaluated the effect of vasopressors to restore blood pressure but can decrease cardiac output (increasing afterload), especially if with coexisting hypovolemia. Vasodilators increase cardiac output, especially in conditions with impaired contractility. Norepinephrine is preferred vasoconstrictor (α-adrenergic effects with some inotropic effects via its moderate β-adrenergic effects). Phosphodiesterase inhibitors combined inotropic and vasodilating effects and (likely) less increase myocardial oxygen requirements than other inotropic agents. Vasopressin, strong vasoconstrictor, excessive increase in arterial pressure but has detrimental effects on cardiac output and splanchnic perfusion [76]. Recently, levosimendan has been shown to be of benefit in decompensated heart failure to improve cardiac performance and achieve a diuresis [77, 78]. Further studies are needed with this agent to determine if it can be used to ameliorate CRS-5.

Renal Support. Currently there are no specific drug-based interventions for renal dysfunction in sepsis [79]. General supportive measures include avoidance of nephrotoxic agents, maintenance of an adequate perfusion pressure and intervention with dialysis [80]. Several studies have evaluated the role of various vasopressor agents in improving renal function in sepsis [81, 82]. There is no role for dopamine to improve renal hemodynamics/function [81, 83], and there have been limited studies with fenoldopam [84, 85]. Norepinephrine increases systemic blood pressure but decreases renal perfusion in normal conditions; NE increases systemic blood pressure and renal perfusion in endotoxic conditions [86]. Vasopressin (natural antidiuretic effects) paradoxically increases urine output and (?) creatinine clearance in patients with septic shock [87–88]. It is currently unclear what the target systemic and intrarenal blood pressure should be to optimize renal function [89].

Management of Consequences: Involvement of the heart and kidney in septic shock portrays special challenges as increasing requirements for vasopressor and ionotrope support are coupled with oliguric states resulting in fluid accumulation and overload. Diuretics have limited roles [90] and RRT particularly with continuous renal replacement should be considered and implemented early [80]. There is some emerging evidence that implementing ultrafiltration early in the course of septic shock may improve outcomes, however further studies are needed to validate these concepts. Maintaining acid base balance and providing adequate 'space' for nutritional support are additional indications for early intervention with RRT.

Research Needs

More detailed epidemiological data...
- time course of developing CRS before/after ICU
- better definition/staging of epidemiological data, i.e. application of RIFLE/AKIN criteria.

• Mechanistic/animal studies with...
- clinically relevant models

- delineation between effects of systemic insult (sepsis) versus organ cross-talk
- application of definition/staging criteria to better compare animal models.
- More detailed observational studies to identify...
- pathways that are associated with development of CRS 5 (pro-/anti-inflammatory pathways activation/role of RAAS)
- site of injury and its effects/prognostic role (glomerular vs. tubular, RV vs. LV failure).
- Experimental studies to elucidate cross-talk between heart and kidney during catastrophic conditions like sepsis, or less dramatic cases like chemo effecting heart and kidney (adriamycin, cisplatin).
- Elucidating these mechanisms could reveal novel biomarkers and improved therapy CRS-5.
- Diagnostic tests: There is a great need to assess the utility of different biomarkers of cardiac and renal dysfunction individually and in combination to diagnose and monitor the progress of CRS-5. These will likely be different for acute and chronic disorders, e.g. sepsis vs. cirrhosis. Techniques to measure biomarker panels and standardized reporting require further research.

Conclusions

CRS-5 is a complex condition which can result from several different conditions that vary in the time and severity. Consequently, it is challenging to attribute the condition to a single pathophysiological mechanism. The timing and sequence of dysfunction of the heart and kidney are influenced by the state of underlying health and the nature and duration of the precipitating condition.

References

1 Ronco C, McCullough PA, Anker SD, Anand I, Aspromonte N, Bagshaw SM, Bellomo R, Berl T, Bobek I, Cruz DN, Daliento L, Davenport A, Haapio M, Hillege H, House A, Katz NM, Maisel A, Mankad S, Zanco P, Mebazaa A, Palazzuoli A, Ronco F, Shaw A, Sheinfeld G, Soni S, Vescovo G, Zamperetti N, Ponikowski P, Acute Dialysis Quality Initiative (ADQI) Consensus Group: Cardiorenal syndromes: an executive summary from the consensus conference of the Acute Dialysis Quality Initiative (ADQI). Contrib Nephrol. Basel, Karger, 2010, vol 165, pp 54–67.

2 Lundy DJ, Trzeciak S: Microcirculatory dysfunction in sepsis. Crit Care Clin 2009;25: 721–731, viii.

3 Trzeciak S, et al: Early microcirculatory perfusion derangements in patients with severe sepsis and septic shock: relationship to hemodynamics, oxygen transport, and survival. Ann Emerg Med 2007;49:88–98, 98.e1–e2.

4 Sakr Y, et al: Persistent microcirculatory alterations are associated with organ failure and death in patients with septic shock. Crit Care Med 2004;32:1825–1831.

5 Jardin F, et al: Sepsis-related cardiogenic shock. Crit Care Med 1990;18:1055–1060.
6 Parker MM, et al: Profound but reversible myocardial depression in patients with septic shock. Ann Intern Med 1984;100:483–490.
7 Lambermont B, et al: Effects of endotoxic shock on right ventricular systolic function and mechanical efficiency. Cardiovasc Res 2003;59:412–418.
8 Dhainaut JF, et al: Coronary hemodynamics and myocardial metabolism of lactate, free fatty acids, glucose, and ketones in patients with septic shock. Circulation 1987;75:533–541.
9 Niederbichler AD, et al: An essential role for complement C5a in the pathogenesis of septic cardiac dysfunction. J Exp Med 2006; 203:53–61.
10 Kumar A, et al: Tumor necrosis factor-α and interleukin-1β are responsible for in vitro myocardial cell depression induced by human septic shock serum. J Exp Med 1996; 183:949–958.
11 Torre-Amione G, et al: Proinflammatory cytokine levels in patients with depressed left ventricular ejection fraction: a report from the Studies of Left Ventricular Dysfunction (SOLVD). J Am Coll Cardiol 1996;27:1201–1206.
12 Benes J, et al: Searching for mechanisms that matter in early septic acute kidney injury: an experimental study. Crit Care 2011;15:R256.
13 Wan L, et al: Angiotensin II in experimental hyperdynamic sepsis. Crit Care 2009; 13:R190.
14 Bougle A, Duranteau J: Pathophysiology of sepsis-induced acute kidney injury: the role of global renal blood flow and renal vascular resistance. Contrib Nephrol. Basel, Karger, 2011, vol 174, pp 89–97.
15 Lerolle N, et al: Renal failure in septic shock: predictive value of Doppler-based renal arterial resistive index. Intensive Care Med 2006;32:1553–1559.
16 Murugan R, et al: Acute kidney injury in non-severe pneumonia is associated with an increased immune response and lower survival. Kidney Int 2010;77:527–535.
17 Papaioannou VE, et al: Relation of heart rate variability to serum levels of C-reactive protein, interleukin-6 and -10 in patients with sepsis and septic shock. J Crit Care 2009; 24:625.e1–e7.
18 Schmidt H, et al: Autonomic dysfunction predicts both 1- and 2-month mortality in middle-aged patients with multiple organ dysfunction syndrome. Crit Care Med 2008; 36:967–970.
19 Tateishi Y, et al: Depressed heart rate variability is associated with high IL-6 blood level and decline in the blood pressure in septic patients. Shock 2007;28:549–553.
20 Ramchandra R, et al: Septic shock induces distinct changes in sympathetic nerve activity to the heart and kidney in conscious sheep. Am J Physiol Regul Integr Comp Physiol 2009;297:R1247–R1253.
21 Meijvis SC, et al: Prognostic value of serum angiotensin-converting enzyme activity for outcome of community-acquired pneumonia. Clin Chem Lab Med 2011;49:1525–1532.
22 Nakada TA, et al: Association of angiotensin II type 1 receptor-associated protein gene polymorphism with increased mortality in septic shock. Crit Care Med 2011;39:1641–1648.
23 Doerschug KC, et al: Renin-angiotensin system activation correlates with microvascular dysfunction in a prospective cohort study of clinical sepsis. Crit Care 2010;14:R24.
24 Shen L, et al: Losartan prevents sepsis-induced acute lung injury and decreases activation of nuclear factor-κB and mitogen-activated protein kinases. Shock 2009;31: 500–506.
25 Hagiwara S, et al: Effects of an angiotensin-converting enzyme inhibitor on the inflammatory response in in vivo and in vitro models. Crit Care Med 2009;37:626–633.
26 Nitescu N, Grimberg E, Guron G: Low-dose candesartan improves renal blood flow and kidney oxygen tension in rats with endotoxin-induced acute kidney dysfunction. Shock 2008;30:166–172.
27 Mortensen EM, et al: Impact of previous statin and angiotensin II receptor blocker use on mortality in patients hospitalized with sepsis. Pharmacotherapy 2007;27:1619–1626.
28 Soni A, et al: Adrenal insufficiency occurring during septic shock: incidence, outcome, and relationship to peripheral cytokine levels. Am J Med 1995;98:266–271.
29 Annane D, et al: Corticosteroids in the treatment of severe sepsis and septic shock in adults: a systematic review. JAMA 2009;301: 2362–2375.

30 Sligl WI, et al: Safety and efficacy of corticosteroids for the treatment of septic shock: a systematic review and meta-analysis. Clin Infect Dis 2009;49:93–101.

31 Kelly KJ: Distant effects of experimental renal ischemia/reperfusion injury. J Am Soc Nephrol 2003;14:1549–1558.

32 Grams ME, Rabb H: The distant organ effects of acute kidney injury. Kidney Int 2012;81: 942–948.

33 Burchill L, et al: Acute kidney injury in the rat causes cardiac remodelling and increases angiotensin-converting enzyme 2 expression. Exp Physiol 2008;93:622–630.

34 Tokuyama H, et al: Macrophage infiltration and cellular proliferation in the non-ischemic kidney and heart following prolonged unilateral renal ischemia. Nephron Physiol 2007; 106:54–62.

35 Hassoun HT, et al: Kidney ischemia-reperfusion injury induces caspase-dependent pulmonary apoptosis. Am J Physiol Renal Physiol 2009;297:F125–F137.

36 Grigoryev DN, et al: The local and systemic inflammatory transcriptome after acute kidney injury. J Am Soc Nephrol 2008;19:547–558.

37 Feltes CM, et al: Pulmonary endothelial cell activation during experimental acute kidney injury. Shock 2011;36:170–176.

38 Liu M, et al: Acute kidney injury leads to inflammation and functional changes in the brain. J Am Soc Nephrol 2008;19:1360–1370.

39 Celes MR, Prado CM, Rossi MA: Sepsis: going to the heart of the matter. Pathobiology 2013;80:70–86.

40 Doi K, et al: Animal models of sepsis and sepsis-induced kidney injury. J Clin Invest 2009; 119:2868–2878.

41 Chopra M, et al: Modulation of myocardial mitochondrial mechanisms during severe polymicrobial sepsis in the rat. PLoS One 2011;6:e21285.

42 Liu M, et al: Transcription factor Nrf2 is protective during ischemic and nephrotoxic acute kidney injury in mice. Kidney Int 2009; 76:277–285.

43 Stengl M, et al: Reduced L-type calcium current in ventricular myocytes from pigs with hyperdynamic septic shock. Crit Care Med 2010;38:579–587.

44 Good DW, George T, Watts BA, 3rd: Toll-like receptor 2 mediates inhibition of HCO_3^- absorption by bacterial lipoprotein in medullary thick ascending limb. Am J Physiol Renal Physiol 2010;299:F536–F544.

45 Hsiao HW, et al: The decline of autophagy contributes to proximal tubular dysfunction during sepsis. Shock 2012;37:289–296.

46 Schreiber A, et al: Acute endotoxemia in mice induces downregulation of megalin and cubilin in the kidney. Kidney Int 2012;82:53–59.

47 Bouchard J, et al: Fluid accumulation, survival and recovery of kidney function in critically ill patients with acute kidney injury. Kidney Int 2009;76:422–427.

48 American College of Chest Physicians/Society of Critical Care Medicine Consensus Conference: Definitions for sepsis and organ failure and guidelines for the use of innovative therapies in sepsis. Crit Care Med 1992;20:864–874.

49 Reinhart K, et al: New approaches to sepsis: molecular diagnostics and biomarkers. Clin Microbiol Rev 2012;25:609–634.

50 Rudiger A, Singer M: Mechanisms of sepsis-induced cardiac dysfunction. Crit Care Med 2007;35:1599–1608.

51 Dellinger RP, et al: Surviving Sepsis Campaign: international guidelines for management of severe sepsis and septic shock: 2008. Crit Care Med 2008;36:296–327.

52 Vignon P, et al: Hand-held echocardiography with Doppler capability for the assessment of critically-ill patients: is it reliable? Intensive Care Med 2004;30:718–723.

53 Bellomo R, et al: Acute renal failure – definition, outcome measures, animal models, fluid therapy and information technology needs: the Second International Consensus Conference of the Acute Dialysis Quality Initiative (ADQI) Group. Crit Care 2004;8:R204–R212.

54 Mehta RL, et al: Acute Kidney Injury Network: report of an initiative to improve outcomes in acute kidney injury. Crit Care 2007; 11:R31.

55 KDIGO Clinical Practice Guideline for Acute Kidney Injury: Summary of recommendation statements. Kidney Int 2012;suppl 2:8–12.

56 Comnick M, Ishani A: Renal biomarkers of kidney injury in cardiorenal syndrome. Curr Heart Fail Rep 2011;8:99–105.

57 Cruz DN, et al: Role of biomarkers in the diagnosis and management of cardio-renal syndromes. Semin Nephrol 2012;32:79–92.

58 Calvin JE, Driedger AA, Sibbald WJ: An assessment of myocardial function in human sepsis utilizing ECG gated cardiac scintigraphy. Chest 1981;80:579–586.
59 Liu KD, et al: Acute kidney injury in patients with acute lung injury: impact of fluid accumulation on classification of acute kidney injury and associated outcomes. Crit Care Med 2011;39:2665–2671.
60 Macedo E, et al: Fluid accumulation, recognition and staging of acute kidney injury in critically-ill patients. Crit Care 2010;14:R82.
61 Bellomo R, et al: Fluid management in septic acute kidney injury and cardiorenal syndromes. Contrib Nephrol. Basel, Karger, 2010, vol 165, pp 206–218.
62 Boyd JH, et al: Fluid resuscitation in septic shock: a positive fluid balance and elevated central venous pressure are associated with increased mortality. Crit Care Med 2011;39: 259–265.
63 Zarjou A, Agarwal A: Sepsis and acute kidney injury. J Am Soc Nephrol 2011;22:999–1006.
64 De Vriese AS, et al: Cytokine removal during continuous hemofiltration in septic patients. J Am Soc Nephrol 1999;10:846–853.
65 Tapia P, et al: Effectiveness of short-term 6-hour high-volume hemofiltration during refractory severe septic shock. J Trauma Acute Care Surg 2012;72:1228–1238.
66 Honore PM, et al: High-volume hemofiltration in sepsis and SIRS: current concepts and future prospects. Blood Purif 2009;28:1–11.
67 Matson J, Zydney A, Honore PM: Blood filtration: new opportunities and the implications of systems biology. Crit Care Resusc 2004;6:209–217.
68 Uchino S, et al: Cytokine dialysis: an ex vivo study. ASAIO J 2002;48:650–653.
69 Nakada TA, et al: Continuous hemodiafiltration with PMMA hemofilter in the treatment of patients with septic shock. Mol Med 2008;14:257–263.
70 Peng Z, et al: Evaluation of the effects of pulse high-volume hemofiltration in patients with severe sepsis: a preliminary study. Int J Artif Organs 2010;33:505–511.
71 Nakamura M, et al: Treatment of severe sepsis and septic shock by CHDF using a PMMA membrane hemofilter as a cytokine modulator. Contrib Nephrol. Basel, Karger, 2010, vol 166, pp 73–82.
72 Cruz DN, et al: Early use of polymyxin B hemoperfusion in abdominal septic shock: the EUPHAS randomized controlled trial. JAMA 2009;301:2445–2452.
73 Humes HD, et al: A selective cytopheretic inhibitory device to treat the immunological dysregulation of acute and chronic renal failure. Blood Purif 2010;29:183–190.
74 Formica M, et al: Coupled plasma filtration adsorption. Contrib Nephrol. Basel, Karger, 2007, vol 156, pp 405–410.
75 Rimmele T, Kellum JA: Clinical review: blood purification for sepsis. Crit Care 2011;15:205.
76 Faivre V, et al: Cardiac and renal effects of levosimendan, arginine vasopressin, and norepinephrine in lipopolysaccharide-treated rabbits. Anesthesiology 2005;103:514–521.
77 Hou ZQ, et al: Effect of levosimendan on estimated glomerular filtration rate in hospitalized patients with decompensated heart failure and renal dysfunction. Cardiovasc Ther 2012 (E-pub ahead of print).
78 Ristikankare A, et al: Effects of levosimendan on renal function in patients undergoing coronary artery surgery. J Cardiothorac Vasc Anesth 2012;26:591–595.
79 May CN, et al: Novel targets for sepsis-induced kidney injury: the glomerular arterioles and the sympathetic nervous system. Exp Physiol 2012;97:1168–1177.
80 Chou YH, et al: Impact of timing of renal replacement therapy initiation on outcome of septic acute kidney injury. Crit Care 2011; 15:R134.
81 Bellomo R, Wan L, May C: Vasoactive drugs and acute kidney injury. Crit Care Med 2008; 36(suppl):S179–S186.
82 Brienza N, et al: A comparison between fenoldopam and low-dose dopamine in early renal dysfunction of critically ill patients. Crit Care Med 2006;34:707–714.
83 Schmoelz M, et al: Comparison of systemic and renal effects of dopexamine and dopamine in norepinephrine-treated septic shock. J Cardiothorac Vasc Anesth 2006;20: 173–178.
84 Cobas M, et al: Fenoldopam in critically ill patients with early renal dysfunction. A crossover study. Cardiovasc Ther 2011;29: 280–284.
85 Landoni G, et al: Fenoldopam in cardiac surgery-associated acute kidney injury. Int J Artif Organs 2008;31:561.

86 Redfors B, et al: Effects of norepinephrine on renal perfusion, filtration and oxygenation in vasodilatory shock and acute kidney injury. Intensive Care Med 2011;37:60–67.
87 Guzman JA, Rosado AE, Kruse JA: Vasopressin vs. norepinephrine in endotoxic shock: systemic, renal, and splanchnic hemodynamic and oxygen transport effects. J Appl Physiol 2003;95:803–809.
88 Gordon AC, et al: The effects of vasopressin on acute kidney injury in septic shock. Intensive Care Med 2010;36:83–91.
89 Deruddre S, et al: Renal arterial resistance in septic shock: effects of increasing mean arterial pressure with norepinephrine on the renal resistive index assessed with Doppler ultrasonography. Intensive Care Med 2007; 33:1557–1562.
90 Nigwekar SU, Waikar SS: Diuretics in acute kidney injury. Semin Nephrol 2011;31:523–534.
91 Muciño-Bermejo J, Carrillo-Esper R, Uribe M, Méndez-Sánchez N: Acute kidney injury in critically ill cirrhotic patients: a review. Ann Hepatol 2012;11:301–310.
92 Wong F, et al: Working Party proposal for a revised classification system of renal dysfunction in patients with cirrhosis. Gut 2011;60: 702–709.
93 Kashani K, et al: Discovery and validation of cell cycle arrest biomarkers in human acute kidney injury. Crit Care 2013;17:R25.

Ravindra L. Mehta, MD, FACP, FASN, FRCP
UCSD Medical Center 8342
200 W Arbor Drive
San Diego, CA 92103 (USA)
E-Mail rmehta@ucsd.edu

McCullough PA, Kellum JA, Mehta RL, Murray PT, Ronco C (eds): ADQI Consensus on AKI Biomarkers and Cardiorenal Syndromes. Contrib Nephrol. Basel, Karger, 2013, vol 182, pp 195–196
DOI: 10.1159/000350012

Appendix 1: 10th Acute Dialysis Quality Initiative (ADQI) Consensus Group

Co-Chairs

Ravindra L. Mehta, MD, Division of Nephrology and Hypertension, Department of Medicine, University of California San Diego, San Diego, Calif., USA

Patrick Murray, MD, School of Medicine and Medical Science, University College Dublin, Dublin, Ireland

Writing Group Members

Josée Bouchard, MD, Division of Nephrology and Hypertension, Division of Nephrology, Université de Montréal, Montréal, Qué., Canada

Sean M. Bagshaw, MD, Division of Critical Care Medicine, Faculty of Medicine and Dentistry, University of Alberta, Edmonton, Alta., Canada

Raj Chakravarthi, MD, Division of Nephrology, Albany Medical College, Albany, N.Y., USA

Lakhmir S. Chawla, MD, Department of Anesthesiology and Critical Care Medicine and Division of Renal Diseases and Hypertension, Department of Medicine, George Washington University Medical Center, Washington, D.C., USA

Dinna Cruz, MD, Department of Nephrology Dialysis and Transplantation, San Bortolo Hospital, Vicenza, Italy

Salvatore Di Somma, MD, Sant'Andrea Hospital, Second Faculty Medical School, La Sapienza University of Rome, Rome, Italy

Kent Doi, MD, Department of Nephrology and Endocrinology, The University of Tokyo, Tokyo, Japan

Peter Doran, MD, Medizinische Poliklinik, University of Munich, Munich, Germany

Jacques Duranteau, MD, Nephrology Unit, Le Kremlin-Bicêtre, France

Zoltan Endre, MD, Christchurch Kidney Research Group, Department of Medicine, University of Otago, Christchurch, New Zealand

Stuart Goldstein, MD, Center for Acute Care Nephrology, Division of Nephrology and Hypertension, The Heart Institute, Cincinnati Children's Hospital Medical Center, Cincinnati, Ohio, USA

Michael Haase, MD, Department of Nephrology, Hypertension, Diabetes and Endocrinology, Otto von Guericke University, Magdeburg, Germany

Can Ice, MD, Department of Intensive Care Adults, Erasmus MC University Medical Centre Rotterdam, Rotterdam, The Netherlands

Bertrand L. Jaber, MD, Division of Nephrology, Tufts Medical Center, Boston, Mass., USA

John A. Kellum, MD, Clinical Research, Investigation, and Systems Modeling of Acute Illness Center, Department of Critical Care Medicine, University of Pittsburgh, Pittsburgh, Pa., USA

Jay L. Koyner, MD, Section of Nephrology, Department of Medicine, University of Chicago, Pritzker School of Medicine, Chicago, Ill., USA

Andrew Lewington, MD, Leeds Teaching Hospitals, Leeds, UK

Etienne Macedo, MD, Division of Nephrology, University of São Paulo, São Paulo, S.P., Brazil

Alan Maisel, MD, Department of Medicine and Cardiology, San Diego VA Medical Center and University of California, San Diego, Calif., USA

Peter A. McCullough, MD, MPH, St. John Providence Health System, Warren, Mich., Providence Hospitals and Medical Centers, Southfield and Novi, Mich., St. John Macomb Oakland Center, Madison Heights, Mich., St. John Hospital and Medical Center, Detroit, Mich., USA

Shane Meehan, MD, Department of Pathology, Section of Nephrology, University of Chicago, Chicago, Ill., USA

Mark Okusa, MD, University of Virginia School of Medicine, Charlottesville, Va., USA

Claudio Ronco, MD, Department of Nephrology, International Renal Research Institute (IRRIV), San Bortolo Hospital, Vicenza, Italy

Andrew Shaw, MD, Department of Anesthesiology, Duke University Medical Center, Durham, N.C., USA

Edward Siew, MD, Division of Nephrology and Hypertension, Department of Medicine, Vanderbilt University School of Medicine, Nashville, Tenn., USA

Ravi Thadhani, MD, Division of Nephrology, Massachusetts General Hospital, Boston, Mass., USA

Sushrut S. Waikar, MD, Renal Division, Department of Medicine, Brigham and Women's Hospital, Harvard Medical School, Boston, Mass., USA

Li Yang, MD, Renal Division, Brigham and Women's Hospital, Boston, Mass., USA

McCullough PA, Kellum JA, Mehta RL, Murray PT, Ronco C (eds): ADQI Consensus on AKI Biomarkers and Cardiorenal Syndromes. Contrib Nephrol. Basel, Karger, 2013, vol 182, pp 197–198
DOI: 10.1159/000350014

Appendix 2: 11th Acute Dialysis Quality Initiative (ADQI) Consensus Group

Co-Chairs

Peter A. McCullough, MD, MPH, St. John Providence Health System, Warren, Mich., Providence Hospitals and Medical Centers, Southfield and Novi, Mich., St. John Macomb Oakland Center, Madison Heights, Mich., St. John Hospital and Medical Center, Detroit, Mich., USA

John A. Kellum, MD, Clinical Research, Investigation, and Systems Modeling of Acute Illness Center, Department of Critical Care Medicine, University of Pittsburgh, Pittsburgh, Pa., USA

Claudio Ronco, MD, Department of Nephrology, International Renal Research Institute (IRRIV), San Bortolo Hospital, Vicenza, Italy

Writing Group Members

Michael Haase, MD, Department of Nephrology, Hypertension, Diabetes and Endocrinology, Otto von Guericke University, Magdeburg, Germany

Christian Müller, MD, University Hospital, Basel, Switzerland

Kevin Damman, MD, Department of Cardiology, University Medical Center Groningen, Groningen, The Netherlands

Patrick Murray, MD, School of Medicine and Medical Science, University College Dublin, Dublin, Ireland

Dinna Cruz, MD, Department of Nephrology Dialysis and Transplantation, San Bortolo Hospital, Vicenza, Italy

Andrew A. House, MD, Schulich School of Medicine and Dentistry, University of Western Ontario Division of Nephrology, University Hospital, London, Ont., Canada

Kai M. Schmidt-Ott, MD, Experimental and Clinical Research Center, Charité-Universitätsmedizin Berlin, Max Delbrück Center for Molecular Medicine, Berlin, Germany

Giorgio Vescovo, MD, Department of Internal Medicine, S. Bortolo Hospital, Vicenza, Italy

Sean M. Bagshaw, MD, Division of Critical Care Medicine, Faculty of Medicine and Dentistry, University of Alberta, Edmonton, Alta., Canada

Eric A. Hoste, MD, Department of Intensive Care Medicine, Ghent University Hospital, Ghent, Belgium

Carlos Briguori, MD, Department of Cardiology, Clinica Mediterranea, Naples, Italy

Branko Braam, MD, Department of Medicine/Division of Nephrology and Immunology, University of Alberta Hospital, Edmonton, Alta., Canada

Lakhmir S. Chawla, MD, Department of Anesthesiology and Critical Care Medicine and Division of Renal Diseases and Hypertension, Department of Medicine, George Washington University Medical Center, Washington, D.C., USA

Maria R. Costanzo, MD, Midwest Heart Foundation, Edward Heart Hospital, Naperville, IL, USA

James A. Tumlin, MD, Department of Medicine, College of Medicine, University of Tennessee, Chattanooga, Tenn., USA

Charles A. Herzog, MD, Hennepin County Medical Center, Minneapolis, Minn., USA

Ravindra L. Mehta, MD, Division of Nephrology and Hypertension, Department of Medicine, University of California San Diego, San Diego, Calif., USA

Hamid Rabb, MD, Nephrology Division, Johns Hopkins Hospital, and the Department of Medicine, Johns Hopkins University School of Medicine, Baltimore, Md., USA

Andrew Shaw, MD, Department of Anesthesiology, Duke University Medical Center, Durham, N.C., USA

Kai Singbartl, MD, Department of Critical Care Medicine, Clinical Research, Investigation, and Systems Modeling of Acute Illness Center, University of Pittsburgh, Pittsburgh, Pa., USA

Author Index

Subject Index